# SpringerBriefs in Mathematics

**SpringerBriefs in Mathematics** showcases expositions in all areas of mathematics and applied mathematics. Manuscripts presenting new results or a single new result in a classical field, new field, or an emerging topic, applications, or bridges between new results and already published works, are encouraged. The series is intended for mathematicians and applied mathematicians.

# BCAM SpringerBriefs



BCAM *SpringerBriefs* aims to publish contributions in the following disciplines: Applied Mathematics, Finance, Statistics and Computer Science. BCAM has appointed an Editorial Board, who evaluate and review proposals.

Typical topics include: a timely report of state-of-the-art analytical techniques, bridge between new research results published in journal articles and a contextual literature review, a snapshot of a hot or emerging topic, a presentation of core concepts that students must understand in order to make independent contributions.

Please submit your proposal to the Editorial Board or to Francesca Bonadei, Executive Editor Mathematics, Statistics, and Engineering: francesca.bonadei@springer.com

More information about this series at http://www.springer.com/series/10030

Debora Amadori · Laurent Gosse

# Error Estimates for Well-Balanced Schemes on Simple Balance Laws

## One-Dimensional Position-Dependent Models

Debora Amadori
DISIM
Università degli Studi dell'Aquila
L'Aquila
Italy

Laurent Gosse
Istituto per le Applicazioni del Calcolo
CNR
Rome
Italy

ISSN 2191-8198 ISSN 2191-8201 (electronic)
SpringerBriefs in Mathematics
ISBN 978-3-319-24784-7 ISBN 978-3-319-24785-4 (eBook)
DOI 10.1007/978-3-319-24785-4

Library of Congress Control Number: 2015953000

Springer Cham Heidelberg New York Dordrecht London


Printed on acid-free paper

Springer International Publishing AG Switzerland is part of Springer Science+Business Media
(www.springer.com)

*To our parents*
*To Sonia*

# Foreword

Well-balanced schemes were introduced in the 1990s for solution of hyperbolic conservation laws with source terms. At that time, the idea of taking into account the source term in the numerical fluxes was very new, and several important issues were not understood. The first one was how to build well-balanced schemes. Indeed, the very nonlinear notion of well-balancing, together with the difficulty of consistency, makes it nontrivial to produce such a scheme, even for simple equations. The second issue was what properties are desirable for well-balanced schemes in order to achieve the best compromise between stability and accuracy. During the 2000s, many methods and ideas offered improvements addressing both these issues, thereby generating a collection of practically efficient schemes.

However, third important question remained essentially open until recently: How to devise methods to rigorously analyze these well-balanced schemes? This is, of course, of key importance in order to understand the limitations of known techniques and to improve them further, in particular when resonance occurs. In this monograph, the authors provide a self-contained exposition of useful tools related to this less well understood issue, including their contributions and most recent achievements. Schemes for both scalar laws and semilinear systems, with position-dependent source terms, are analyzed in the spirit of Glimm, with augmented Riemann problems and Lyapunov functionals. Error estimates are established, and a particular form of these estimates concerning the growth in time and the rates in terms of space and time increments offers perhaps the most important characterization of well-balancing that is available at the level of numerical analysis. An exploratory two-dimensional study is also provided, which raises delicate questions. All of the material provided in this book is highly relevant for the understanding of well-balanced schemes and will contribute to future improvements.

Marne-la-Vallee
May 2015

François Bouchut

# Preface

The present book originates both from the talks delivered by the first author at several international conferences and from a mini-course given by the second author at BCAM in November 2014. The scope is narrower compared with its companion reference,[1] as most of the aspects related to linear (or weakly nonlinear) kinetic equations have been omitted in order to focus on the rigorous derivation of global error estimates for particular types of (systems of) balance laws in one space dimension.

The monograph presents, in a hopefully attractive and self-contained form, some techniques based on the $L^1$ stability theory derived at the end of the 1990s by A. Bressan, T.-P. Liu, and T. Yang, which yield original error estimates for so-called well-balanced numerical schemes solving one-dimensional hyperbolic systems of balance laws.[2] Efforts have been focused on a practical assessment of these error bounds, too, either by a wave-front tracking technique or by a simpler Godunov process.

Well-balanced schemes, as they are studied hereafter, mostly rely on a reformulation of the original balance laws as a homogeneous, nonconservative system involving one supplementary steady "fake variable" often denoted $a(x)$. In a strictly hyperbolic regime, a scattering state emerges from the time decay of an extended interaction potential, including the "standing waves" associated with $a$. Such an asymptotic picture motivates a treatment of source terms, originally suggested by James Glimm,[3] such as "local scattering centers", which we shall apply extensively.

We warmly thank Prof. Enrique Zuazua, who encouraged us to write this manuscript and to submit it for publication in the BCAM Springer Briefs collection.

L'Aquila  
July 2015

Debora Amadori  
Laurent Gosse

---

[1] *Computing Qualitatively Correct Approximations of Balance Laws*, Springer (2013).

[2] *cf.* Marc Laforest, *SIAM J. Math. Anal.* **35** (2004), 1347–1370.

[3] *cf.* J. Glimm and D.H. Sharp, *Found. Phys.* **16** (1986), 125–141.

# Contents

# Acronyms

| | |
|---|---|
| *BV* | Banach space of functions with bounded variation |
| CFL | Courant–Friedrichs–Lewy |
| FD | Finite-Differences |
| FS | Fractional-Step |
| GNL | Genuinely Non-Linear |
| LD | Linearly Degenerate |
| LTE | Local Truncation Error |
| NC | Non-Conservative |
| ODE | Ordinary Differential Equation |
| PDE | Partial Differential Equation |
| RH | Rankine–Hugoniot |
| TS | Time-Splitting |
| TV | Total Variation |
| WFT | Wave-Front Tracking |
| WB | Well-Balanced |

# Chapter 1
# Introduction

**Abstract** In this chapter we introduce the topic of the book, namely the class of partial differential equations on which it is focused and an Outline of the presented material.

**Keywords** Hyperbolic system of balance laws · Accuracy of schemes for balance laws

## 1.1 Some General Perspective

Even if post-WWII supersonic aircrafts development considerably stimulated both the understanding and practical approximation schemes for inviscid fluid mechanics equations, multidimensional systems of conservation laws [9], the decision to massively invest into the computational treatment of Euler equations appears to be a bit more recent, roughly speaking the 80s, when the shortcomings of popular widely-used models based on potential flow hypotheses were revealed during the so-called "Stockholm Olympics", see [29]. A previous success for (hypersonic) computational fluid dynamics was the correct design of the thermal shield on the Apollo spatial vehicles for the atmosphere reentry phase, based on a multi-D extension of the Godunov scheme produced in 1959 for inviscid Euler equations, [14, 35]. Oppositely, during the 80s, the Airbus 320 transonic airfoils were still designed by means of a potential flow approximation, i.e. a scalar equation for which general conservation of mass, momentum and energy is not completely ensured.

### 1.1.1 Inviscid Systems of Conservation Laws

Obviously, an heavy price inherent to switching from the computational simulation of an elementary, possibly fitted, model to the handling of a considerably more complex system of equations is the design of reliable, realizable and efficient numerical algorithms. Presently, multidimensional inviscid Euler equations appear to be replete



D. Amadori and L. Gosse, *Error Estimates for Well-Balanced Schemes on Simple Balance Laws*, SpringerBriefs in Mathematics,
DOI 10.1007/978-3-319-24785-4_1

of difficulty of all kinds, like the mathematical ones in order to give some rigorous sense to weak discontinuous solutions containing shock waves, [5, 26], but also some unexpected ones, like several non-uniqueness and ill-posedness phenomena recently established for 2D inviscid systems containing vorticity, [8, 12]. It came with no surprise that fine-scale structures concomitant with vorticity developing in the solution may result into a destabilizing factor in numerical computations; less expected was the fact that it overcomes the stabilizing effect of entropy dissipation, even at the mathematical level of continuous equations. Paradoxically, some authors begin to take a step back and advocate some potential flow models: see [11].

Besides, left apart ill-posedness pathologies, numerical issues may pop up immediately when one plans to compute multidimensional flows endowed with discontinuous shock waves by means of finite-differences on a fixed grid: P.L. Roe calls the example of a static shear flow not aligned with the grid (i.e. an oblique contact discontinuity) the "*case against upwinding*": see [19, p. 345] and Fig. 6.7. Examples of this kind sparkled many developments and algorithmic advances, some of which were consigned in the book [20], see also [36]. Even if many pathologies still remain in existing algorithms, some signs suggest that algorithmic innovation slowed down so much that it practically came to a stop, [31], the main driving force behind recent computational progress being mostly an increase in CPU power.

### *1.1.2 One-Dimensional Systems*

Enormous difficulties coming from the multidimensional characters of compressible flows [13] can be somewhat circumvented by following Godunov's ideas, namely narrowing the scope only to one-dimensional models within a dimensional-splitting perspective (see e.g. [17, 24]). *Necessity knows no law.* A salient feature in one-dimensional inviscid models is the availability of an efficient tool for the resolution of discontinuities, the Riemann problem and its "simple waves" endowed with neat directions of propagation; such a building-block gave rises to many time-marching approximation procedures, which all rely on resolving discontinuities (possibly induced through wave interactions) by correctly distinguishing the nature of left- and right-going disturbances. A piecewise-constant approximation of initial data furnishes a convenient input allowing to set up many Riemann-based algorithms, leading to significant theoretical stability results, too [5]. Of course, one should always keep in mind that, apart from specific problems, one-dimensional models are a drastic simplification which cannot be thought as truly reflecting the complexity of certain multidimensional inviscid flows [10, 16, 27, 30, 40].

Even in such a simplified context, both theoretical and computational issues remain: left aside the scalar conservation law for which a satisfying theory was proposed by Kružkov [23] and the so-called Temple class systems [6], general 1D strictly hyperbolic systems of conservation laws are usually subject to restrictive smallness assumptions on their (initial, boundary) data [39]. In the realm of such a

$L^1$-stability theory of (small) $BV$ entropy solutions, one still faces several important obstructions when considering numerical approximation processes, except for a few notable cases, like Godunov's scheme for Temple class systems [7] (in contrast with [2, 3]), or the wavefront tracking algorithm [28], a convenient convergence theory of widely-used schemes is still not available. Worse, obstacles appear to be not just technical, as some quite practical examples were produced, displaying for instance wave-curve deformations as a result of numerical viscosity [1, 22, 37].

### *1.1.3 Source Terms Inclusion*

So far, only homogeneous models were tacitly considered, that is to say, models endowed with neither forcing nor dissipation, for which global conservation exactly holds along time. Needless to say, a good description of realistic problems often involves supplementary, lower-order terms, usually referred to as *source terms*, likely to restrict, or at least perturb a bit more existing stability theories. Two main classes of source terms can be distinguished:

- Dissipative mechanisms, like relaxation appearing in discrete models of more complex kinetic equations, which effect is to push the system onto an *equilibrium manifold* where its dynamics are described by a reduced number of variables.
- Possibly accretive terms of bounded extent thus naturally position-dependent, yielding a *scattering mechanism* as they are somewhat negligible at infinity, but create strong interactions with convective waves in the vicinity of the origin.

Numerical techniques for both these categories are intrinsically different: uniform in space, dissipative terms are well handled by operator-splitting in time algorithms, consisting essentially in alternating the treatment of the homogeneous equations and the ordinary differential system resulting of the presence of the source (which acts more like a sink) [17]. Oppositely, position-dependent, bounded-extent source terms lead, in large times, to a scattering (i.e. non-interacting) state where homogeneous hyperbolic waves propagate far away, but a steady-state wave, resulting from a delicate balancing of convection forces and sources, stands close to the origin. Such a balance is usually not correctly reproduced by a time-splitting strategy, as was already well understood in the 80s [18, 34]. A scattering mechanism is present in homogeneous systems of conservation laws: by strict hyperbolicity, the solution decays toward a set of non-interacting waves. This set of waves identifies with the Riemann problem at infinity, obtained by a "zoom-out" rescaling. Yet, in presence of a bounded extent source, such a rescaling wipes off everything except for a Dirac measure in zero inside which all the effects are concentrated. A $S$-matrix relates incoming and outgoing hyperbolic waves, locally scattered by a "Dirac source" [13].

## 1.2 This Book in a Nutshell

### *1.2.1 Outline of the Contents*

This book is mostly concerned with quantifying the accuracy of specific numerical approximations of 1D systems of balance laws endowed with position-dependent, possibly accretive source terms, hence decaying in large time onto a scattering state involving a stationary wave. These approximations follow Glimm's canvas, namely reducing a position-dependent source term to a countable collection of "local scattering centers" by means of Dirac measures (for instance, located at the edges of computational cells in a Cartesian mesh), and proceeding by iteratively solving all the resulting Riemann problems, but this time, including the "standing wave" (in the terminology of [21]) which results of the "Dirac source scattering". Within a Godunov scheme, this procedure is mostly referred to as *well-balanced scheme* [15] whereas in the context of homogenization of periodically forced scalar laws, one speaks more willingly about a *generalized Glimm scheme* [38]. During the last decade, such a numerical strategy was very successful on practical computations for shallow water equations endowed with steep topography source terms (see [4] and (2.11); hence the natural question "*is there a rigorous mathematical explanation for this class of schemes outperforming the standard ones ?*" to which the forthcoming chapters aim at (partly) answer in a unified, self-contained manner.

- Chapter 2 deals with several forms of error estimates for 1D systems of balance laws. Especially, the classical so-called *Local Truncation Error (L.T.E.)* is introduced, following an analysis of K. Morton. Most "high-order schemes" for conservation laws rely on this (formal) notion of local error for their construction. Since it usually employs Taylor expansions of the solution, its relevance in the context of weak, possibly discontinuous, solutions isn't obvious: two simple examples (the deformation of shock curves by numerical viscosity and a nonlinear shock-rarefaction interaction) reveal some of its shortcomings. Moreover, such a local error bound should be integrated in time, with the help of convenient uniform bounds, in order to produce a global error estimate which quantifies the actual error separating the approximate solution from its exact counterpart in some norm (usually $L^1$): by assuming infinite smoothness for solutions of a linearized shallow water system, it is shown that handling the topography term $a(x)$ as a collection of "local scattering centers" furnishes a much more robust approximation. This elementary calculation is checked on a practical example.
- Chapter 3 focuses onto the 1D, possibly accretive, position-dependent scalar balance law (3.1). For simplicity, only one numerical algorithm is considered, the wavefront-tracking (WFT) [5, 10], for which a general $L^1$ stability theory was established by means of a specific Lyapunov functional. Such a functional keeps track of the time-evolution of the distance between two approximate WFT solutions emerging from different $BV$ initial data. It's easy to see that under some mild hypotheses (one of these being the non-resonance assumption [21, 25]), this

functional being uniformly equivalent to the $L^1$ norm furnishes a reliable error estimate *allowing to bypass the Gronwall lemma*, hence free from any exponential dependence in time! Another feature of such an analysis is revealed when considering the particular case of a periodically forced scalar law (3.42), for which an original variant of the stability functional allows to improve significantly previous estimates. Practical tests are displayed for both cases, too.

- Chapter 4 is probably the most ambitious; it explores the possibility to extend the good error estimates of the former one, from the scalar equation toward a semilinear, position-dependent system endowed with a relaxation-type source term. Systems of this type can be met in many areas of application, left aside well-known relaxation approximations to scalar conservation laws. Being semilinear, it allows for the derivation of an original, time-decaying, Lyapunov functional which uniformly controls the global $L^1$ error of the numerical process. Such a functional is absolutely specific to the "local scattering centers" procedure and it's rather easy to produce examples for which one sees without doubt that it outperforms the Kuznetsov-type estimates which govern the error of more standard discretizations. This fact is checked numerically, too, in Chap. 5.
- Last but not least, Chap. 6 aims at giving hints about more complex models for which our analysis may be extended: such systems may include weakly nonlinear models involving a coupling with a self-consistent Poisson equation (rendering electrostatic interactions, biological confinement or gravity forces) or so-called Temple class systems met for instance in the context of traffic flow modeling. A widely open question remains 2D applications: as Chap. 4 was largely concerned with a semilinear problems being lower-order perturbations of the 1D wave equation, the Riemann problem for the 2D linear wave equation in variables $p, u, v$ is studied: an analytical expression of its exact solution is given and displayed on Fig. 6.5. Very complex wave interactions appear inside the Kirchhoff disc as soon as initial data are endowed with some vorticity. This is the basic building block in order to construct a genuinely 2D Godunov scheme.

### 1.2.2 The Ariadne's Thread

Basically, our conclusions can be stated in a straightforward fashion as follows:

1. Glimm's method of concentrating bounded extent, position-dependent source terms as "local scattering centers" arranged on a discrete lattice consists essentially in treating them by means of supplementary wave interactions.
2. If the source term reads $k(x)g(u)$, it is convenient to rewrite it $g(u)\partial_x a$, being $a(x)$ an antiderivative of $k$ satisfying the trivial equation $\partial_t a = 0$. One immediately deduces that $u, a$ solve an homogeneous system, with a supplementary (immobile) characteristic family, the "standing wave" in [21] (see also [25, 33]).
3. Under the non-resonance assumption, such an augmented system is strictly hyperbolic hence admits a Lyapunov functional decaying along convenient pair of *BV*

solutions, say $u, a$ and $v, b$. This decay is proved without invoking the Gronwall lemma, so no time-exponentials are involved. The source term's effects are generally handled by means of an interaction potential, a classical object in Glimm's theory which may impose smallness restrictions, though.
4. Being that Lyapunov functional uniformly equivalent with the $L^1$ norm, a global error estimate is easily deduced. However, since it depends only on the $BV$ norms of the data, mechanically it perceives only the $L^1$ norm of the position-dependent coefficient. For instance, the presence of $k(x)g(u)$ produces an error depending only of $\|k\|_{L^1}$, but not on any of its derivatives. *This is most probably the origin of the accuracy on shallow water models endowed with a steeply varying topography term*: the global error are insensitive to its oscillations (such a remark is relevant in homogenization of oscillating balance laws [38], too).
5. The non-resonance assumption is not to be taken lightly: a formal analysis of constants showing up in the error estimate for balance laws reveals a term like $|g(u)/f'(u)|$, being $f'(u)$ the velocity field. Such a quantity blows up if the augmented system ceases to be strictly hyperbolic [21], so one must expect the accuracy to decrease in the vicinity of sonic or stagnation points. See for instance Fig. 3.3, where an "error spike" is located at a point where $f'(u) \simeq 0$.

## References

1. M. Arora, P.L. Roe, On postshock oscillations due to capturing schemes in unsteady flows. J. Comput. Phys. **130**, 25–40 (1997)
2. P. Baiti, A. Bressan, H.K. Jenssen, An instability of the Godunov scheme. Commun. Pure Appl. Math. **59**, 1604–1638 (2006)
3. S. Bianchini, $BV$ solutions of the semidiscrete upwind scheme. Arch. Rational Mech. Anal. **167**, 1–81 (2003)
4. F. Bouchut, *Nonlinear Stability of Finite Volume Methods for Hyperbolic Conservation Laws, and Well-balanced Schemes for Sources*. Frontiers in Mathematics Series (Birkhäuser, Basel, 2004). ISBN 3-7643-6665-6
5. A. Bressan, *Hyperbolic Systems of Conservation Laws—The One-Dimensional Cauchy Problem*. Oxford Lecture Series in Mathematics and its Applications, vol. 20 (Oxford University Press, Oxford, 2000)
6. A. Bressan, P. Goatin, Stability of $L^\infty$ solutions of Temple class systems. Differ. Integral Equ. **13**, 1503–1528 (2000)
7. A. Bressan, H.K. Jenssen, On the convergence of Godunov scheme for straight line nonlinear hyperbolic systems. Chin. Ann. Math. (CAM) **21**, 269–284 (2000)
8. E. Chiodaroli, A counterexample to well-posedness of entropy solutions to the compressible Euler system. J. Hyperbolic Differ. Equ. **11**(2014), 493–519 (2014)
9. R. Courant, K.O. Friedrichs, *Supersonic Flow and Shock Waves* (Springer, New York, 1976)
10. C.M. Dafermos, *Hyperbolic Conservation Laws in Continuum Physics*, 3rd edn. (Springer, Heidelberg, 2010)
11. V. Elling, Relative entropy and compressible potential flow. Acta Math. Sci. Ser. B Engl. Ed. **35**(4), 763–776 (2015)
12. V. Elling, The carbuncle phenomenon is incurable. Acta Math. Scientia **29B**, 1647–1656 (2009)
13. J. Glimm, D.H. Sharp, An $S$-matrix theory for classical nonlinear physics. Found. Phys. **16**, 125–141 (1986)

14. S.K. Godunov, , V.S. Ryabenkii, Difference Schemes; An Introduction to the Underlying Theory, Studies in Mathematics and Its Applications **19**, North Holland (1987)
15. J.M. Greenberg, A.Y. LeRoux, A well-balanced scheme for the numerical processing of source terms in hyperbolic equations. SIAM J. Numer. Anal. **33**, 1–16 (1996)
16. J. Guckenheimer, Shocks and rarefactions in two space dimensions. Arch. Rational Mech. Anal. **59**, 281–291 (1975)
17. H. Holden, K.H. Karlsen, K.-A. Lie, N.H. Risebro, *Splitting Methods for Partial Differential Equations with Rough Solutions. Analysis and MATLAB Programs.* Series of Lectures in Mathematics (European Mathematical Society (EMS), Zürich, 2010)
18. L. Huang, T.P. Liu, A conservative, piecewise-steady difference scheme for transonic nozzle flow. Comput. Math. Appl. **12A**, 377–388 (1986)
19. M.Y. Hussaini, A. Kumar, M.D. Salas (eds.), *Algorithmic Trends in Computational Fluid Dynamics (English)* (Springer, Berlin, 1993)
20. M.Y. Hussaini, B. van Leer, J. Van Rosendale (eds.), *Upwind and High-Resolution Schemes (English)* (Springer, Berlin, 1997)
21. E. Isaacson, B. Temple, Convergence of the $2 \times 2$ Godunov method for a general resonant nonlinear balance law. SIAM J. Appl. Math. **55**, 625–640 (1995)
22. S. Jin, J.-G. Liu, The effects of numerical viscosities. I. Slowly moving shocks. J. Comput. Phys. **126**, 373–389 (1996)
23. S.N. Kružkov, First order quasilinear equations with several independent variables. Math. Sbornik (N.S.) **81**(123), 228255 (1970)
24. R.J. LeVeque, *Numerical Methods for Conservation Laws.* ETH Zurich (Birkhauser, Basel, 1992)
25. C. Li, T.P. Liu, Asymptotic states for hyperbolic conservation laws with a moving source. Adv. Appl. Math. **4**, 353–379 (1983)
26. A. Majda, *Compressible Flow in Several Space Variables* (Springer, New York, 1984)
27. J. Rauch, BV estimates fail for most quasilinear hyperbolic systems in dimensions greater than one. Commun. Math. Phys. **106**, 481–484 (1986)
28. N.H. Risebro, A front tracking alternative to the random choice method. Proc. Am. Math. Soc. **117**, 1125–1139 (1993)
29. A. Rizzi, H. Viviand (eds.), *Numerical Methods for the Computation of Inviscid Transonic Flows with Shock Waves.* Notes on Numerical Fluid Mechanics Series, vol. 3 (Vieweg Verlag, Braunschweig, 1981)
30. P.L. Roe, Discrete models for the numerical analysis of time-dependent multidimensional gas dynamics. J. Comput. Phys. **63**, 458 (1986)
31. P.L. Roe, Computational fluid dynamics—retrospective and prospective. Int. J. Comput. Fluid Dyn. **19**(8), 581–594 (2005)
32. B. Sjögreen, http://www.math.fsu.edu/~sussman/Bjorn_Sjogreen_Notes.pdf
33. G.A. Sod, A random choice method with application to reaction-diffusion systems in combustion. Comput. Math. Appl. **11**, 129–144 (1985)
34. G.A. Sod, A numerical study of oxygen diffusion in a spherical cell with the Michaelis-Menten oxygen uptake kinetics. J. Math. Biol. **24**, 279–289 (1986)
35. T.D. Taylor, B.S. Masson, Application of the unsteady numerical method of Godunov to computation of supersonic flows past bell shaped bodies. J. Comput. Phys. **5**, 443–454 (1970)
36. E.F. Toro, *Riemann Solvers and Numerical Methods for Fluid Dynamics: A Practical Introduction*, 3rd edn. (Springer, Berlin, 2009)
37. J. Von Neumann, R.D. Richtmyer, A method for the numerical calculation of hydrodynamic shocks. J. Appl. Phys. **21**, 232–237 (1950)
38. E. Weinan, Homogenization of scalar conservation laws with oscillatory forcing terms. SIAM J. Appl. Math. **52**, 959–972 (1992)
39. R. Young, *On elementary interactions for hyperbolic conservation laws.* Unpublished (1993). http://www.math.umass.edu/~young/Research/misc/elem.pdf
40. Y. Zheng, *Systems of Conservation Laws: Two-Dimensional Riemann Problems.* Progress in Nonlinear Differential Equations and Their Applications, vol. 38 (Birkhauser Verlag, Boston, 2001)

# Chapter 2
# Local and Global Error Estimates

**Abstract** In this chapter we analyze some simple examples, which suggest that the error quantification should take into account of the possible grow in time of the error. This observation provides a motivation for going beyond more classical local-in-time concepts of error (so-called *Local Truncation Error*).

**Keywords** Accuracy of schemes for balance laws · Local truncation error

It's a well-known fact that quasilinear hyperbolic equations generally admit only weak solutions, in the sense that discontinuities develop and propagate along distinguished directions (at least in one space dimension, the situation in 2D being more delicate). Their mathematical analysis is usually carried out within a Banach space of discontinuous functions with finite total variation, $BV(\mathbb{R})$, sometimes intersected with $L^1(\mathbb{R})$, in order to ensure some integrability properties (positive total mass should be preserved).

However, when it comes to derive numerical algorithms meant to approximate their (entropy) solutions on a countable lattice, people frequently evoke so-called "high-order schemes", which may seem a puzzling notion, as they rely on Taylor expansions. Besides, Kuznetsov's method indicates that, as soon as numerical viscosity is present in the numerical process, the best convergence rate (hence an upper bound for the $L^1$ global error) is of the order of $\sqrt{t \cdot \Delta x}$, much less than any high-order rate. So, in which sense should one understand high-order, left apart the fact that one cannot reach more than second-order accuracy when computing only local cell averages?

Hereafter, important differences between local truncation errors, a notion inherited from smooth solutions of Ordinary Differential Equations (ODE), and global errors for possibly weak solutions of Partial Differential Equations (PDE) is put at forefront. It is explained that "high-order", as a notion, essentially refers to a local in time residual, in vicinity of a point where the considered solution is as smooth as possible, which should be further bounded uniformly and integrated in order to produce a meaningful global error estimate. This sheds light on the apparent discrepancy between second-order, so-called MinMod schemes (see e.g. [3, 27, 28]) for linear advection, and the actual convergence rates (slightly over $\frac{1}{2}$, see [19]) that were obtained only recently. Local estimates are most of the times insensitive to



D. Amadori and L. Gosse, *Error Estimates for Well-Balanced Schemes on Simple Balance Laws*, SpringerBriefs in Mathematics,
DOI 10.1007/978-3-319-24785-4_2

specific features of elaborate numerical schemes [9, 11], whereas global ones may be better suited for such purposes. This is illustrated on a model of linearized shallow water equations over topography, admitting smooth solutions.

## 2.1 Notion of Local Truncation Error (LTE)

Let's focus onto a one-dimensional convex scalar conservation law,

$$\partial_t u + \partial_x f(u) = 0, \qquad u(t=0,\cdot) = u_0 \in BV(\mathbb{R}), \qquad (t,x) \in \mathbb{R}_*^+ \times \mathbb{R}, \quad (2.1)$$

within a Cartesian uniform computational grid, determined by a space-step $\Delta x$ and a time-step $\Delta t$ satisfying the standard homogeneous CFL restriction. By defining $C_k = (x_{k-\frac{1}{2}}, x_{k+\frac{1}{2}})$ as the generic computational cell of width $\Delta x$ centered on $x_k = k\Delta x$, $k \in \mathbb{Z}$, one may apply the Divergence Theorem on any rectangle $C_k^n := C_k \times (t^n, t^{n+1})$ in order to derive a mass-preserving numerical scheme for (2.1):

$$\int_{C_k} u\left(t^{n+1}, x\right) \mathrm{d}x = \int_{C_k} u\left(t^n, x\right) \mathrm{d}x - \int_{t^n}^{t^{n+1}} f\left(u\left(\tau, x_{k+\frac{1}{2}}\right)\right) - f\left(u\left(\tau, x_{k-\frac{1}{2}}\right)\right) \mathrm{d}\tau.$$

This is equivalent to the weak formulation of (2.1) with test-functions being indicator functions of $C_k$, denoted $\chi(C_k)$. Hereafter, as $u_k^n = \int_{C_k} u(t^n, x)\frac{\mathrm{d}x}{\Delta x}$, the observation yielding Godunov's scheme is the following: in case $u(t^n, \cdot)$ is constant on each computational cell $C_k$, then the boundary flux terms can be explicitly obtained by resolving all the discontinuities, that is to say, Riemann problems at both interfaces $x_{k\pm\frac{1}{2}}$. Moreover, since Riemann fans $\omega(\frac{x}{t}; u^L, u^R)$ display a self-similar structure,

$$\int_{t^n}^{t^{n+1}} f\left(u\left(\tau, x_{k+\frac{1}{2}}\right)\right) \mathrm{d}\tau = \Delta t \cdot f\left(\omega\left(0; u_k^n, u_{k+1}^n\right)\right). \quad (2.2)$$

In the context of an explicit time-marching algorithm, one may want to get rid of the Riemann solution $\omega$, thus (2.2) rewrites as a smooth and consistent *numerical flux function*, $F : \mathbb{R}^2 \to \mathbb{R}$,

$$\forall u, v \in \mathbb{R}^2, \qquad F(u,v) = f\left(\omega(0; u, v)\right) = \begin{cases} \min\limits_{u \le \xi \le v} f(\xi) & \text{if } u \le v \\ \max\limits_{v \le \xi \le u} f(\xi) & \text{if } u > v. \end{cases} \quad (2.3)$$

For any indexes $k, n \in \mathbb{Z} \times \mathbb{N}$, the Godunov averaging furnished an approximate (formally first-order) value $u_k^n \simeq u(t^n, x_k)$. In order to locally increase its accuracy, a piecewise-linear reconstruction can be set up in each cell,

$$v_k^n : C_k \to \mathbb{R}, \qquad v_k^n(x) = u_k^n + (x - x_k)\sigma_k^n. \quad (2.4)$$

A common way to proceed is by analogy with Lax-Wendroff second-order scheme for (2.1) with linear flux, $f(u) = u$: the resulting definition of local slopes reads,

$$\sigma_k^n = \frac{u_{k+1}^n - u_k^n}{\Delta x}\phi(r_k^n), \qquad r_k^n = \frac{u_k^n - u_{k-1}^n}{u_{k+1}^n - u_k^n}.$$

The slope-limiter $\phi$ must meet with several constraints in order to ensure both the TVD property and formal second-order accuracy in smooth regions, see [26].

*Remark 2.1* Another way to motivate MUSCL piecewise-linear reconstructions is to work out the ODE system obtained by semi-discretization in space (the "Method of Lines", evoked in [15]) in order to obtain a Local (space-) Truncation Error in $\Delta x^2$ for smooth exact solutions $u$: see e.g. [20, 21, 29, 30].

### 2.1.1 Semi-discretization in Space (Method of Lines)

Hereafter we follow the canvas of Cullen and Morton [6] in order to shed some light onto various general mechanisms of error creation/propagation (see also [4, 30]). Let a Cauchy problem for a given partial differential operator $\mathscr{L}$ be,

$$\partial_t u = \mathscr{L} u, \qquad u(t = 0, \cdot) = u_0. \tag{2.5}$$

For $\Delta x > 0$ fixed and the corresponding griding of the real line, a finite-differences approximation of $\mathscr{L}$ acting on $\Delta x \cdot \mathbb{Z}$ is denoted by $\mathscr{L}_{\Delta x}$, so (2.5) reduces to an (infinite) differential system (*Method of Lines*, in [15, Chap. 17]), with $\tilde{u}(t, \cdot) \in \ell^\infty(\mathbb{Z})$:

$$\frac{\mathrm{d}}{\mathrm{d}t}\tilde{u} = \mathscr{L}_{\Delta x}\tilde{u}, \qquad \tilde{u}(t = 0, \cdot) = \mathscr{P}_{\Delta x} u_0, \tag{2.6}$$

for which one can worry about the global error $u - \tilde{u}$ at each time $t > 0$.

- One "triangulates" $u(t, \cdot) - \tilde{u}(t, \cdot)$ by inserting $\mathscr{P}_{\Delta x} u(t, \cdot)$,

$$u - \tilde{u} = (Id - \mathscr{P}_{\Delta x})u + (\mathscr{P}_{\Delta x} u - \tilde{u}) := a_{\Delta x} + e_{\Delta x},$$

  where $a_{\Delta x}$ is purely an approximation error, but $e_{\Delta x}$ stands for an evolutionary error, which may accumulate in time, and satisfies a differential equation,

$$\frac{\mathrm{d}}{\mathrm{d}t} e_{\Delta x} = \frac{\mathrm{d}}{\mathrm{d}t}\mathscr{P}_{\Delta x} u - \frac{\mathrm{d}}{\mathrm{d}t}\tilde{u} = \mathscr{P}_{\Delta x}\mathscr{L} u - \mathscr{L}_{\Delta x}\tilde{u}. \tag{2.7}$$

- Triangulating again, one gets

$$\frac{\mathrm{d}e_{\Delta x}}{\mathrm{d}t} = (\mathscr{P}_{\Delta x}\mathscr{L} u - \mathscr{L}_{\Delta x}\mathscr{P}_{\Delta x} u) + (\mathscr{L}_{\Delta x}\mathscr{P}_{\Delta x} u - \mathscr{L}_{\Delta x}\tilde{u}),$$

leading to,

$$\frac{\mathrm{d}}{\mathrm{d}t} e_{\Delta x} + (\mathscr{L}_{\Delta x} \tilde{u} - \mathscr{L}_{\Delta x} \mathscr{P}_{\Delta x} u) = (\mathscr{P}_{\Delta x} \mathscr{L} u - \mathscr{L}_{\Delta x} \mathscr{P}_{\Delta x} u) := L.T.E.,$$

and by substituting $\tilde{u}$ by $\mathscr{P}_{\Delta x} u - e_{\Delta x}$, we get finally:

$$\boxed{\frac{\mathrm{d}}{\mathrm{d}t} e_{\Delta x} + \left[\mathscr{L}_{\Delta x}(\mathscr{P}_{\Delta x} u - e_{\Delta x}) - \mathscr{L}_{\Delta x} \mathscr{P}_{\Delta x} u\right] = L.T.E.,} \tag{2.8}$$

(L.T.E. standing for Local Truncation Error). Hence, the L.T.E. is just a source term inside the differential equation (2.8) governing the scheme's evolutionary error; this was noted in [17, 20, 30], too.

In case both (2.5) and its (consistent) discrete approximation $\mathscr{L}_{\Delta x}$, are dissipative ("contractive" [21, 29], "strongly stable" in a terminology of [17]) in some norm, this source term is responsible for most of the error $e_{\Delta x}$; if, on the contrary, (2.5) happens to be accretive, for instance if $\|u(t) - v(t)\| \leq K \|u_0 - v_0\|$ with $K > 1$ like in Bressan-Glimm's theory of strictly hyperbolic systems of conservation laws [5], then both $\mathscr{L}_{\Delta x}$ and the L.T.E. can contribute to the increase of the evolutionary error.

*Remark 2.2* If the approximation $\mathscr{L}_{\Delta x}$ is linear, then (2.8) simplifies into,

$$\boxed{\forall t > 0, \qquad \frac{\mathrm{d}}{\mathrm{d}t} e_{\Delta x}(t) = \mathscr{L}_{\Delta x} e_{\Delta x}(t) + \tau_u(t),}$$

where $\tau_u(t)$ stands for the L.T.E. related to ($x$-derivatives of) the exact solution $u(t, \cdot)$ to (2.5) at time $t$. Duhamel's principle yields an expression of the evolutionary error,

$$e_{\Delta x}(t) = \exp(t \cdot \mathscr{L}_{\Delta x}) \left( e_{\Delta x}(t = 0) + \int_0^t \exp(-s \cdot \mathscr{L}_{\Delta x}) \tau_u(s) \mathrm{d}s \right).$$

Quantities like $\exp(t \cdot \mathscr{L}_{\Delta x})$ may be estimated by "logarithmic norms", see e.g. [21].

The local truncation error is intrinsically a "low frequency" information: being a byproduct of a Taylor expansion, it tacitly assumes that higher-order derivatives of the solution get smaller. Accordingly, it was recast in the formalism of Fourier analysis in, e.g., [31, sect. 2.4]. In order to cope with oscillating, possibly high-frequency, solutions of linear transport (advection) equations, finite-difference schemes endowed with "spectral-like resolution" were devised in [12, 14, 25] for which both numerical dissipation and dispersion were carefully scrutinized. For specific problems, like the linear wave equation recast as a multi-D "div-grad" hyperbolic system, numerical schemes can be derived from the exact Kirchoff's method of spherical means: see [1] and especially the connections with the classical Lax-Wendroff formalism.

### 2.1.2 Local Truncation Error (LTE) and Second-Order Accuracy

Second-order accuracy in space for (2.1), or linear advection equations, was studied in [18] (see also [10, 27]). These equations are dissipative in $L^1$, so the former analysis yielding (2.8) indicates that the local truncation error is probably the main source of evolutionary error. For $\mathscr{L}u = -\partial_x f(u)$, it reads: $\forall k \in \mathbb{Z}$,

$$\mathscr{P}_{\Delta x}\mathscr{L}u(t,x_k) = -\frac{1}{\Delta x}\int_{x_{k-\frac{1}{2}}}^{x_{k+\frac{1}{2}}} \partial_x f(u)\mathrm{d}x = -\frac{f\left(u\left(t,x_{k+\frac{1}{2}}\right)\right) - f\left(u\left(t,x_{k-\frac{1}{2}}\right)\right)}{\Delta x},$$

by exact integration of the conservation law (2.1). Now, since high-order accuracy is only concerned with smooth exact solutions $u$, one approximates this expression with a second-order mid-point rule by taking advantage of $x_{k+\frac{1}{2}} = \frac{x_{k+1}+x_k}{2}$,

$$\mathscr{P}_{\Delta x}\mathscr{L}u(t,x_k) = \frac{f\left(\frac{u(t,x_{k+1})+u(t,x_k)}{2}\right) - f\left(\frac{u(t,x_k)+u(t,x_{k-1})}{2}\right) + O(\Delta x^2)}{\Delta x},$$

and so, the L.T.E. is the difference between this approximation and the numerical scheme $\mathscr{L}_{\Delta x}$ applied to the piecewise constant projection of the exact solution, $\mathscr{P}_{\Delta x}u$. Since $\mathscr{L}_{\Delta x}$ needs to be conservative and consistent with $\mathscr{L}$, we assume it is given by a (smooth) numerical flux which reads, in standard notation,

$$\tilde{F}_{k+\frac{1}{2}} = F\left(u^L_{k+\frac{1}{2}}, u^R_{k+\frac{1}{2}}\right), \qquad \mathscr{L}_{\Delta x}\mathscr{P}_{\Delta x}u(t,x_k) = \frac{\tilde{F}_{k+\frac{1}{2}}(t) - \tilde{F}_{k-\frac{1}{2}}(t)}{\Delta x},$$

where $u^{L/R}_{k+\frac{1}{2}}$ are obtained from the set of cell-centered values $\mathscr{P}_{\Delta x}u$ by means of a reconstruction like (2.4) and $F$ is, for instance, the exact Godunov flux (2.3). Hence,

$$\boxed{L.T.E. = \frac{\left[f\left(\frac{u(t,x_{k+1})+u(t,x_k)}{2}\right) - \tilde{F}_{k+\frac{1}{2}}\right] - \left[f\left(\frac{u(t,x_k)+u(t,x_{k-1})}{2}\right) - \tilde{F}_{k-\frac{1}{2}}\right]}{\Delta x}.}$$

As the CFL condition imposes $\Delta t = O(\Delta x)$, second-order accuracy asks for,

$$\left| f\left(\frac{u(t,x_{k+1})+u(t,x_k)}{2}\right) - \tilde{F}_{k+\frac{1}{2}}(t)\right| = O(\Delta x^2),$$

which, by the smoothness of the flux functions, reduces simply to,

$$\forall t,k \in \mathbb{R}^+ \times \mathbb{Z}, \qquad \left| u^{L/R}_{k+\frac{1}{2}}(t) - \frac{u(t,x_{k+1})+u(t,x_k)}{2}\right| = O(\Delta x^2). \tag{2.9}$$

And this meets with the definition used by Osher (see Lemma 2.1, in [18, p. 953]) and Sjogreen (see Theorem 3.9 in [24, p. 47]). A slightly different derivation of a second-order scheme for smooth solutions is given in [4, p. 53], essentially by keeping the term $\frac{\mathrm{d}}{\mathrm{d}t}\mathscr{P}_{\Delta x}u$ in (2.7) inside the expression of the L.T.E as follows:

$$\begin{aligned}\tfrac{\mathrm{d}}{\mathrm{d}t}\mathscr{P}_{\Delta x}u(t,\cdot) &= \lim_{\Delta t\to 0}\left(\tfrac{\mathscr{P}_{\Delta x}u(t+\Delta t,\cdot)-\mathscr{P}_{\Delta x}u(t,\cdot)}{\Delta t}\right)\\ &= -\frac{F\big(u(t,\cdot+\Delta x),u(t,\cdot)\big)-F\big(u(t,\cdot),u(t,\cdot-\Delta x)\big)}{\Delta x},\end{aligned}$$

where $F$ is the exact flux defined in (2.3). The L.T.E. is now defined like,

$$\boxed{\forall k\in\mathbb{Z},\qquad \frac{\mathrm{d}}{\mathrm{d}t}\mathscr{P}_{\Delta x}u(t,x_k)-\mathscr{L}_{\Delta x}\mathscr{P}_{\Delta x}u(t,x_k) = -\frac{\mathscr{F}_{k+\frac{1}{2}}(t)-\mathscr{F}_{k-\frac{1}{2}}(t)}{\Delta x},}$$

where $\mathscr{F}_{k+\frac{1}{2}}(t) = F\big(u(t,x_k+\Delta x),u(t,x_k)\big)-F_{k+\frac{1}{2}}(t)$. The scheme induced by the numerical flux $F_{k+\frac{1}{2}}$ is called second-order in space as soon as, for any smooth exact solution $u(t,\cdot)$, $\mathscr{F}_{k+\frac{1}{2}}$ is a quadratic quantity (possibly depending on $|\partial_{xx}u(t,\cdot)|$),

$$\forall t\geq 0,\qquad |\mathscr{F}_{k+\frac{1}{2}}(t)| = O(\Delta x^2). \tag{2.10}$$

Clearly, both criteria pick up variants of the (unstable) "centered scheme" which is second-order, but unstable because it lets the total variation increase strongly.

### *2.1.3 Illustration of Two Errors Not Controlled by LTE*

The L.T.E. was defined as, given an approximate function space, the difference between the projection of the exact (smooth) solution and the outcome of the numerical process, generally after an infinitesimally small time-step $\Delta t > 0$, with identical initial values. Its shortcomings are henceforth revealed by putting in default these two main assumptions: smoothness and identical data.

Let's first present a classical example of issues which result from a lack of smoothness in the exact solution of the considered problem. Assume (2.5) is a $2\times 2$ genuinely nonlinear system of conservation laws, like for instance the $p$-system. If prescribed initial data produces (after some time) a forward-going 2-shock, the numerical viscosity inherent in a standard Godunov-type scheme (2.6) breaks that shock into smaller ones, and the resulting discontinuity appears to be spread on a certain number of computational cells: see Fig. 2.1. However, the 2-shock present in the exact solution is meant to connect two states $(p_L, u_L)$ and $(p_R, u_R)$ belonging to the 2-shock curve in phase space, which isn't a straight line because the considered system isn't in the Temple class. As soon as it gets broken into smaller jumps by means of artificial viscosity, there's no reason for those ones to belong to the same identical curve: hence, as the time-marching scheme will advance by solving elementary Riemann

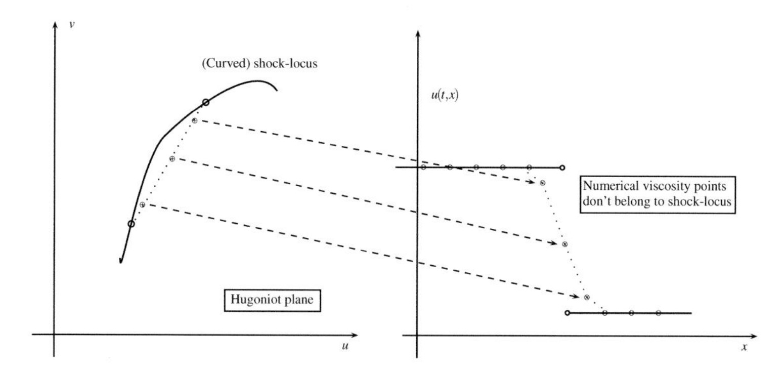


**Fig. 2.1** Numerical viscosity and its effects for systems with curved shock loci

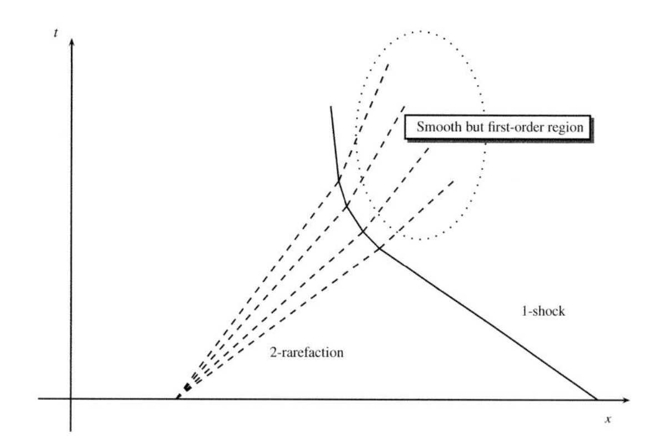


**Fig. 2.2** Nonlinear $2 \times 2$ interaction and loss of accuracy in a smooth region

problems, spurious 1-waves are going to develop inside the numerical shock layer, as shown in e.g. [13]. Formal second-order schemes improve marginally the situation by displaying shocks spread on a reduced area, but can't eliminate this drawback of shock-capturing techniques. A thorough analysis of the deformation of shock curves in phase space was achieved by [2] and later in [23].

The second well-known example of a fallout for L.T.E. consists in setting up initial data for which there's no approximation error, for instance two Riemann problems exactly represented on the computational grid, but yielding two waves of different nature which interact in finite time. For instance, always considering the $p$-system, suppose that in $x = -1$ one prescribes a jump between $(p_L, u_L)$ and $(p_M, u_M)$ belonging to the 2-rarefaction curve, whereas in $x = 1$, another one between $(p_M, u_M)$ and $(p_R, u_R)$ belonging to the 1-shock curve. After some time, these two approaching waves interact and a new middle state $(\tilde{p}_M, \tilde{u}_M)$ appears with the 1-shock on the left and the 2-rarefaction on its right: see Fig. 2.2. This pattern is called a "scattering state", meaning that no wave interaction can occur anymore. Incoming waves are both the smooth 2-rarefaction, inside which a second-order scheme will actually display its formal accuracy, and the 1-shock, in the vicinity of which only first-order accuracy holds for the sake of stability. The 2-rarefaction spawn from the interaction may not be second-order accurate with respect to the exact solution because it results of the mixing of both first and second order data. This example was first presented in [8], and then studied in [7, 22] (see also [16, 32]).

## 2.2 Linearized Shallow Water with Topography

Hereafter, an elementary example, despite its simplicity, will reveal already a part of the specific features that govern the more complex non-linear cases when source terms are included into a system of equations. By linearizing around a static state $\bar{\rho} > 0$, $\bar{u} = 0$ the usual one-dimensional shallow water equations with topography,

$$\partial_t \rho + \partial_x(\rho u) = 0, \qquad \partial_t(\rho u) + \partial_x\left(\rho u^2 + \frac{\rho^2}{2}\right) = -\rho\partial_x a, \tag{2.11}$$

the following system arises,

$$\partial_t \rho + \partial_x J = 0, \qquad \partial_t J + \partial_x \rho = -\partial_x a \tag{2.12}$$

where $a = a(x)$ stands for the bottom. By linearity, solutions $\rho(t,\cdot)$, $J(t,\cdot)$ display identical integrability and smoothness as initial data. An alternative, customary way to deal with system (2.12) is to consider $a = a(x)$ as an independent variable,

$$\begin{cases} \partial_t \rho + \partial_x J = 0, \\ \partial_t J + \partial_x(\rho + a) = 0 \\ \partial_t a = 0. \end{cases} \tag{2.13}$$

### 2.2.1 *Study of the Error Growth in Time*

Characteristic speeds of system (2.12) are $\pm 1$; diagonal variables $f^{\pm} = \rho \pm J$ satisfy

$$(\partial_t - \partial_x) f^- = \partial_x a, \qquad (\partial_t + \partial_x) f^+ = -\partial_x a; \tag{2.14}$$

equivalently, the system (2.13) is diagonalized as follows:

$$\begin{cases} \partial_t(f^{\pm} + a) \pm \partial_x(f^{\pm} + a) = 0 \\ \partial_t a = 0. \end{cases} \tag{2.15}$$

Accordingly, two approaches coexist for approximating the solutions of (2.12):

1. The standard "centered source method" which consists in processing a set of two advection equations with a source term, (2.14).
2. The "well-balanced method", which treats two *homogeneous* advection equations from (2.15): $\partial_t f^\pm \pm \partial_x(f^\pm + a) = 0$.

Define the Courant number $\nu$ as $\Delta t = \nu \Delta x$ with $0 < \nu \leq 1$, together with the function $a \in C^\infty(\mathbb{R})$ having compact support, and $f^\pm(t = 0, \cdot) \in C^3 \cap W^{3,\infty}(\mathbb{R})$. With respect to the preceding section, errors in both space and time will be considered.

### 2.2.2 Analysis of Scheme 1

In each $C_j^n$, the residual $R_j^n$ of the "centered source method" is computed by plugging exact solutions into each inhomogeneous advection equation: for instance,

$$\frac{f^+(t^{n+1}, x_j) - f^+(t^n, x_j)}{\Delta t} + \frac{f^+(t^n, x_j) - f^+(t^n, x_{j-1})}{\Delta x} + k(x_j) = R_j^n,$$

which is the L.T.E. of the scheme in both space and time. By Taylor expansion,

$$|R_j^n| \leq \frac{\Delta x}{2}(1 - \nu)\, \|\partial_{xx} f^\pm(t^n, \cdot)\|_\infty + \frac{\Delta t}{2} \|\partial_{xx} a(x)\|_\infty + C(\Delta x)^2$$

where the $C$ above depends on both $\|\partial_{xxx} f^\pm(t^n, \cdot)\|_\infty$ and $\|\partial_{ttt} f^\pm(t^n, \cdot)\|_\infty$. By linearity, the second derivative above, $\partial_{xx} f^\pm(t, \cdot)$, is bounded by

$$\|\partial_{xx} f^\pm(t, \cdot)\|_\infty \leq \|\partial_{xx} f^\pm(t = 0)\|_\infty + t\|a\|_{C^3}.$$

There are two error amplification mechanisms for the "centered source method": the fact that $R_j^n$ contains $\Delta t \|\partial_{xx} a\|_\infty$ which doesn't vanish as $\nu = \frac{\Delta t}{\Delta x} = 1$ (numerical viscosity), and the linear growth of $t \mapsto \|\partial_{xx} f^\pm(t, \cdot)\|_\infty$. Yet, the pointwise error,

$$(E^\pm)_j^n = (f^\pm - f^\pm_{\Delta x})(t^n, x_j),$$

(where $f^\pm_{\Delta x}$ stands for the piecewise-constant approximation of the exact solution $f^\pm$) satisfies a slightly modified upwind scheme that easily rewrites under the form of a convex combination plus a source term: for instance,

$$(E^+)_j^{n+1} = (E^+)_j^n \left(1 - \frac{\Delta t}{\Delta x}\right) + \frac{\Delta t}{\Delta x}(E^+)_{j-1}^n + \Delta t R_j^n.$$

It remains to take the modulus, maximize, and to sum over $n$ all the residuals in order to derive that, for any $t \geq 0$, the following error estimate holds:

$$\|E^{\pm}(t,\cdot)\|_{L^\infty} \leq \|E^{\pm}(t=0,\cdot)\|_{L^\infty}$$
$$+ \Delta x \cdot t\Big[\nu\|a\|_{C^2} + (1-\nu)\left(\|f^{\pm}(t=0,\cdot)\|_{C^2} + t\|a\|_{C^3}\right)\Big] + O(1)t(\Delta x)^2 \tag{2.16}$$

where the $O(1)$ above depends on $\|f^{\pm}(t=0,\cdot)\|_{C^3}$, $\|a\|_{C^4}$ and linearly on $t$.

### 2.2.3 Analysis of Scheme 2

Oppositely, for the well-balanced method, the same LTE analysis starts with:

$$\frac{f^+(t^{n+1},x_j) - f^+(t^n,x_j)}{\Delta t} + \frac{(f^+ + a)(t^n,x_j) - (f^+ + a)(t^n,x_{j-1})}{\Delta x} = \tilde{R}^n_j.$$

The full upwinding of the topography function $a$ yields a smaller residual:

$$|\tilde{R}^n_j| \leq \frac{\Delta x}{2}(1-\nu)\left\|\partial_{xx}[f^{\pm}(t,\cdot) + a(\cdot)]\right\|_\infty + C(\Delta x)^2,$$

with $C$ depending on the sup norm of $\partial_{xxx}(f^{\pm} + a)$. Since diagonal variables $V = (f^{\pm} + a, a)$ are constant along their characteristics, the sup-norm of their derivatives doesn't grow in time:

$$|\tilde{R}^n_j| \leq \frac{\Delta x}{2}(1-\nu)\left\|\partial_{xx}[f^{\pm}(t=0,\cdot) + a(\cdot)]\right\|_\infty + C_0(\Delta x)^2,$$

for some constant $C_0$. Again, the scheme governing the pointwise error

$$(\tilde{E}^{\pm})^n_j = [(f^{\pm} + a) - (f^{\pm}_{\Delta x} + a)](t^n,x_j)$$

rewrites as a convex combination, so a better global error estimate follows:

$$\|\tilde{E}^{\pm}(t,\cdot)\|_{L^\infty} \leq \|\tilde{E}^{\pm}(t=0,\cdot)\|_{L^\infty}$$
$$+ \Delta x \cdot \frac{t}{2}(1-\nu)\|f^{\pm}(t=0,\cdot) + a\|_{C^2} + C_0 t(\Delta x)^2. \tag{2.17}$$

### 2.2.4 Summary and Illustration

Let's enumerate the main differences between (2.16) and (2.17):

1. The error (2.16) of the standard method is a polynomial (quadratic in the present case) function of time whereas the well-balanced estimate (2.17) is linear.

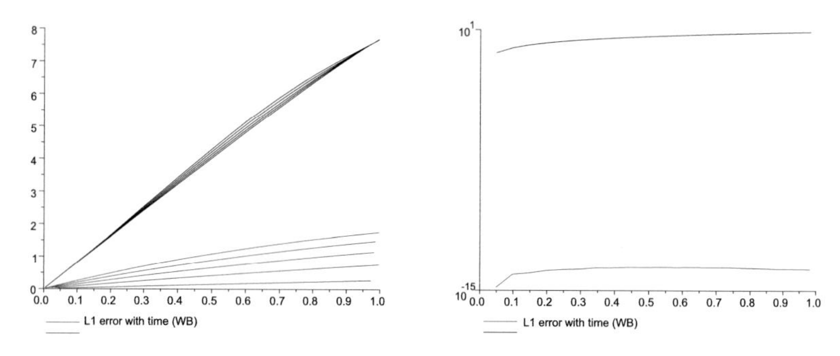


**Fig. 2.3** Time-evolution of the $L^\infty$ errors of well-balanced (*blue*) and centered source (*black*) methods for smooth $f^\pm, a$: $\nu = 0.9, 0.7, 0.5, 0.3, 0.1$ (*left*) and $\nu = 1$ (*right*)

2. In (2.16), the source term always contributes at the same rate regardless to the value of the Courant number $\nu$. Oppositely, fixing $\nu = 1$ implies that for the well-balanced method, the initial error (2.17) stands still,

$$\|\tilde{E}^\pm(t,\cdot)\|_{L^\infty} \simeq \|\tilde{E}^\pm(t=0,\cdot)\|_{L^\infty}.$$

3. The dependence on $a$ and its derivates of the two estimates (2.16) and (2.17) is much different: for small times (at the first order in $t$), (2.16) depends on

$$\nu\|a\|_{C^2} + (1-\nu)\,\|f^\pm(t=0,\cdot)\|_{C^2}$$

while (2.17) depends on

$$(1-\nu)\|f^\pm(t=0,\cdot)+a\|_{C^2}.$$

That is, compensation between $a$ and the initial data is allowed. Moreover, as observed before, the quantity in the above formula vanishes for $\nu = 1$.

On Fig. 2.3, both these theoretical aspects are illustrated: the graphic on the left shows that the well-balanced pointwise error $\tilde{E}$ grows more strongly when $\nu \to 0$ (the numerical viscosity increases) and the right one reveals that $\tilde{E}$ remains constant in time when $\nu = 1$ but this nice property doesn't hold for a conventional method.

*Remark 2.3* One may wonder if the global error estimate for the time-splitting strategy may be improved, in particular its dependence with respect to the derivatives of $a(x)$. A simple computation, relying on elementary stability theory of position-dependent ODE, reveals that in general this cannot be the case: the space dependence of $a$ (more precisely, on $k = a'$) usually introduces additional variation in the unknown, when the time-split update is applied. Intuitively, this occurs because the differential equation in time, at each time-step, is sensitive to the changing values of $k$. Let $g : \mathbb{R} \to \mathbb{R}$ be bounded and Lipschitz continuous and pick $k_1, k_2 \in \mathbb{R}^2$, corresponding solutions $y_1(t)$, $y_2(t)$

$$\dot{y}_1 = k_1 \cdot g(y_1), \qquad \dot{y}_2 = k_2 \cdot g(y_2), \qquad y_1(0) = y_2(0),$$

clearly satisfy[1]

$$|y_1(t) - y_2(t)| \leq \|g\|_\infty |k_2 - k_1| t. \tag{2.18}$$

For a scalar law with source term $\partial_t u + \partial_x f(u) = k(x)g(u)$, see next Chap. 3, a time splitting approach typically asks for solving $\partial_t u = k(x)g(u)$; from (2.18), a local amplification of the approximation $u^{\Delta x}$, of the order of $t \cdot \text{TV}(k)$, is induced. Most of the available error estimates for such equations contain $\text{TV}(u^{\Delta x})$, thus one may expect global errors of time-splitting strategies to grow with $t \cdot \text{TV}(k)$, especially if $g'(u)$ is nonnegative. Oppositely, well-balanced strategies can be proved to obey error bounds depending only on the $L^1$ norm of $k$: assuming that $k \in L^1 \cap BV(\mathbb{R})$ has compact support, say $(a, b)$, there exists an optimal constant of Poincaré's inequality,

$$\boxed{\|k\|_{L^1(a,b)} \leq \frac{b-a}{2} \text{TV}(k),}$$

a quite interesting property for applications where $k$ displays steep gradients, like for instance (2.11) with a topography term, or periodic forcing as well.

## 2.3 Preliminary Conclusions

The L.T.E., as recalled in the beginning of this chapter, is too weak as a notion to reliably quantify the errors generated by a numerical scheme approximating one-dimensional (systems of) balanced because it doesn't take into account for several things, like for instance the time-propagation of incomplete initial data approximations. More specifically, the loss of smoothness in a compressive shock can induce peculiar issues in the form of spurious waves of different characteristic families, which may react in the presence of an accretive source term. Those issues are still not perceived by the L.T.E. which generally isn't well suited for dealing with accretive problems. Error growth taking place in nonlinear wave interactions displays similar features, namely it combines local reduction of accuracy close to a shock wave with data mixing. It therefore follows that a more relevant object of interest appears to be the global error, considered as a function depending on both the time and grid-size. Especially, it perceives a possible time-amplification of errors already present in the approximation of initial data. Moreover, it is able to fully account for limited smoothness of solutions in case it doesn't rely on Taylor expansions. This is the main tool which is about to be exploited in the sequel of this book.

[1] Indeed, let $Y(t)$ satisfy $Y' = g(Y)$, $Y(0) = y_1(0) = y_2(0)$, so that $y_j(t) = Y(k_j t)$, $j = 1, 2$, and then

$$|y_1(t) - y_2(t)| = \left| \int_{k_2 t}^{k_1 t} g(Y(\tau))\, d\tau \right| \leq \|g\|_\infty |k_2 - k_1| t.$$

# References

1. B. Alpert, L. Greengard, T. Hagstrom, An integral evolution formula for the wave equation. J. Comput. Phys. **162**(2), 536–543 (2000)
2. M. Arora, P.L. Roe, On postshock oscillations due to capturing schemes in unsteady flows. J. Comput. Phys. **130**, 25–40 (1997)
3. C. Berthon, C. Sarazin, R. Turpault, Space-time generalized Riemann problem solvers of order $k$ for linear advection with unrestricted time step. J. Sci. Comput. **55**, 268–308 (2013)
4. F. Bouchut, *Nonlinear stability of finite volume methods for hyperbolic conservation laws, and well-balanced schemes for sources*, Frontiers in Mathematics Series (Birkhäuser, Boston, 2004). ISBN 3-7643-6665-6
5. A. Bressan, *Hyperbolic systems of conservation laws—the one-dimensional Cauchy problem*, vol. 20, Oxford Lecture Series in Mathematics and Its Applications (Oxford University Press, Oxford, 2000)
6. M.J.P. Cullen, K.W. Morton, Analysis of evolutionary error in finite element and other methods. J. Comput. Phys. **34**, 245–267 (1980)
7. G. Efraimsson, G. Kreiss, A remark on numerical errors downstream of slightly viscous shocks. SIAM J. Numer. Anal. **36**, 853–863 (1999)
8. B. Engquist, B. Sjögreen, The convergence rate of finite difference schemes in the presence of shocks. SIAM J. Numer. Anal. **35**, 2464–2485 (1998)
9. H. Gilquin, Une famille de schémas numériques T.V.D. pour les lois de conservation hyperboliques. RAIRO—Model. Math. Anal. Numer. **20**, 429–460 (1986)
10. J.B. Goodman, R.J. LeVeque, A geometric approach to high resolution TVD schemes. SIAM J. Numer. Anal. **25**, 268–284 (1988)
11. L. Gosse, MUSCL reconstruction and Haar wavelets. Commun. Math. Sci. **13**, 1501–1514 (2015)
12. Z. Haras, S. Ta'asan, Finite difference schemes for long-time integration. J. Comput. Phys. **114**(2), 265–279 (1994)
13. S. Jin, J.-G. Liu, The effects of numerical viscosities. I. Slowly moving shocks. J. Comput. Phys. **126**, 373–389 (1996)
14. S. Lele, Compact finite difference schemes with spectral-like resolution. J. Comput. Phys. **103**(1), 16–42 (1992)
15. R.J. LeVeque, *Numerical Methods for Conservation Laws* (Birkhauser, ETH Zurich, Basel, 1992)
16. R. Menikoff, Errors when shock waves interact due to numerical shock width. SIAM J. Sci. Comput. **15**, 1227–1242 (1994)
17. K.W. Morton, On the analysis of finite volume methods for evolutionary problems. SIAM J. Numer. Anal. **35**, 2195–2222 (1998)
18. S. Osher, Convergence of generalized MUSCL schemes. SIAM J. Numer. Anal. **22**, 947–961 (1985)
19. B. Popov, O. Trifonov, Order of convergence of second order schemes based on the MINMOD limiter. Math. Comput. **75**, 1735–1753 (2006)
20. J.M. Sanz-Serna, J.G. Verwer, Convergence analysis of one-step schemes in the method of lines. Appl. Math. Comput. **31**, 183–196 (1989)
21. J.M. Sanz-Serna, J.G. Verwer, Stability and convergence at the PDE/stiff ODE interface. Appl. Numer. Math. **5**, 117–132 (1989)
22. M. Siklosi, G. Efraimsson, Analysis of first order errors in shock calculations in two space dimensions. SIAM J. Numer. Anal. **43**, 672–685 (2005)
23. M. Siklosi, B. Batzorig, G. Kreiss, An investigation of the internal structure of shock profiles for shock capturing schemes. J. Comput. Appl. Math. **201**, 8–29 (2007)
24. Sjögreen, B.: Lecture notes. http://www.math.fsu.edu/~sussman/Bjorn_Sjogreen_Notes.pdf
25. B. Swartz, B. Wendroff, The relative efficiency of finite difference and finite element methods. I: hyperbolic problems and splines, SIAM J. Numer. Anal. **11**, 979–993 (1974)

26. P.K. Sweby, High resolution schemes using flux limiters for hyperbolic conservation laws. SIAM J. Numer. Anal. **21**, 995–1011 (1984)
27. E.F. Toro, *Riemann Solvers and Numerical Methods for Fluid Dynamics: A Practical Introduction*, 3rd edn. (Springer, New York, 2009)
28. B. Van Leer, Towards the ultimate conservative difference scheme, V. A second order sequel to Godunov's method. J. Comput. Phys. **32**, 101–136 (1979)
29. J.G. Verwer, Contractivity in locally one-dimensional splitting methods. Numer. Math. **44**, 247–259 (1984)
30. J.G. Verwer, J.M. Sanz-Serna, Convergence of method of lines approximations to partial differential equations. Computing **33**, 297–313 (1984)
31. R. Vichnevetsky, J.B. Bowles, Fourier analysis of numerical approximations of hyperbolic equations. SIAM J. Applied Math. **5** (1982)
32. Zaide, D.W.-M.: Numerical shock-wave anomalies. Ph.D. thesis, University of Michigan (2012)

# Chapter 3
# Position-Dependent Scalar Balance Laws

**Abstract** In this chapter we illustrate our approach to the error estimate analysis, for a scalar, space-dependent, non-resonant balance law. A wave-front tracking scheme is analyzed, leading to a generic linear dependence in time of the error. Numerical illustrations are given for accretive case and for the case of periodic forcing.

**Keywords** Space-dependent balance laws · Non-resonance

Simple error bounds presented in former Sect. 2.2 suggest that handling source terms at interfaces, that is to say by means of supplementary wave interactions, brings potentially more involved Riemann solvers, but this is balanced by the following rigorous benefits in terms of robustness and accuracy:

- The growth of the global error (expressed in a strong norm, like for instance $L^1$) with respect to time is much slower, especially in the presence of an accretive source term.
- In this global error, there is no dependence in high order derivatives in $x$ of the position-dependent coefficient, which is especially relevant in the context of 1D shallow water models endowed with sharply-varying topography terms.

The present chapter is meant to rigorously establish both these properties in the context of 1D scalar balance laws. For the equation $\partial_t u + \partial_x f(u) = k(x)g(u)$, Kružkov's theory provides an estimate on the $L^1$ distance between the difference of two *exact* solution. Such $L^1$-norm may grow in time at the rate of $\exp(Nt)$, where

$$N = \sup_{x,u} k(x)g'(u).$$

When $N > 0$, we will refer to the case as *accretive* case. Oppositely, when $N$ vanishes or $N < 0$, the $L^1$-norm will be either non-expanding or contracting, respectively. It is reasonable to expect these properties on exact solutions be reflected in the quality of numerical approximations. Following [10], we will focus on two main cases: one is accretive, $N > 0$, while the second concerns a source term being only space-dependent $g' = 0$ hence $N = 0$. In this second case the argument will be adapted in order to cope with periodic forcing. Both cases will be treated under the restriction of



D. Amadori and L. Gosse, *Error Estimates for Well-Balanced Schemes on Simple Balance Laws*, SpringerBriefs in Mathematics,
DOI 10.1007/978-3-319-24785-4_3

"non-resonance": characteristic velocities do not vanish, $f'(u) \neq 0$, which is equivalent to impose that augmented system (3.3) remains always strictly hyperbolic. Strict hyperbolicity is known to imply several types of decay properties in entropy solutions (see also [7]); such features yield stability in $L^1$, and consequently error bounds, without any use of Gronwall's Lemma (which would bring time-exponentials in the picture, a severe drawback for numerical applications [11]).

## 3.1 Non-resonant Wave-Front Tracking Algorithm

To fix ideas, consider the following Cauchy problem

$$\partial_t u + \partial_x f(u) = k(x)g(u), \qquad u(t=0,x) = u_0(x) \in BV(\mathbb{R}) \tag{3.1}$$

for $x \in \mathbb{R}$, under the assumptions

$$f,\ g \in C^2, \qquad \inf_u f'(u) > 0, \qquad k(x) \in L^1_{loc}. \tag{3.2}$$

Setting $a(x) = \int^x k(s)\,ds$ as an indefinite antiderivative of $k$, the Eq. (3.1) rewrites as an elementary Temple class system, with characteristic speeds $0$, $f'(u)$:

$$\partial_t u + \partial_x f(u) - g(u)\partial_x a = 0, \qquad \partial_t a = 0. \tag{3.3}$$

The $2 \times 2$ system (3.3) is naturally endowed with two Riemann invariants. One is the static function $a$, corresponding to the characteristic speed $f'(u)$; along the level sets of $a$, the system reduces to the purely convective equation $\partial_t u + \partial_x f(u) = 0$. On the other hand, the additional field corresponding to the 0 characteristic speed is related to the stationary equation $\partial_x f(u) - g(u)\partial_x a = 0$. Therefore, wherever $g(u) \neq 0$, along steady solutions one has

$$\phi(u) - a = \text{const.}, \quad \phi' = \frac{f'}{g}.$$

Any smooth function of $\phi(u) - a$ is a Riemann invariant, in regions where $g(u) \neq 0$. In order to deal with source term that possibly vanish at isolated points, we introduce the following Riemann invariant

$$\boxed{w(u,a) = \phi^{-1}\big(\phi(u) - a\big).} \tag{3.4}$$

Initial data belongs to an invariant domain for the system (3.3). In terms of Riemann coordinates $(a, w)$, invariant domains correspond simply to rectangles. We define $w_0(x) = w(u_0(x), a(x))$ and assume that the data are confined into a rectangle:

$$(a(x), w_0(x)) \in K \doteq [\bar{a}_1, \bar{a}_2] \times [\bar{w}_1, \bar{w}_2] \tag{3.5}$$

for some constants $\bar{a}_1 < \bar{a}_2, \bar{w}_1 < \bar{w}_2$. This assumption is quite reasonable in several cases; as an example, it is met for *every* bounded initial data $(a(x), u_0(x))$ with $a \in BV(\mathbb{R})$ if we assume that the trajectories of ordinary differential equation

$$\frac{du(a)}{da} = \frac{g(u)}{f'(u)} \tag{3.6}$$

do not blow up in finite intervals (see [2, Eq. (1.3)] and [18]), which holds by (3.7).

### 3.1.1 Properties of the Riemann Invariant $w(u, a)$

The above choice, in presence of isolated zeros of $g$, is convenient since it provides a continuous function $u \to w(u, a)$ across these points. Indeed let us assume, for simplicity, that $g(u)$ has a finite number of zeroes, say $\{\bar{u}_j\}_{j=1 \dots N_0}$ being an increasing sequence; no specific sign is required for $g$ outside the set of zeroes. Again for simplicity we assume that $f'/g$ is not integrable at $u \to \pm\infty$:

$$\int^{\pm\infty} \frac{f'(u)}{g(u)}\, du = \infty. \tag{3.7}$$

Setting $I_j \doteq (\bar{u}_j, \bar{u}_{j+1})$ for $j = 0, \dots, N_0$, with $\bar{u}_0 = -\infty$, $\bar{u}_{N_0+1} = +\infty$, $\phi(u)$ is invertible on each interval $I_j$ with full range $\mathbb{R}$, so that $w(u, a)$ as in (3.4) is well defined for $u \in I_j, a \in \mathbb{R}$. Besides, $w$ extends continuously to isolated zeros of $g$ [8]:

$$\forall\, \bar{u} \in \mathbb{R}, \quad g(\bar{u}) = 0 \quad \Rightarrow \quad \lim_{u \to \bar{u}} w(u, a) = \bar{u}.$$

Indeed, by (3.4), the mean value theorem applies to $\phi(w) - \phi(u) = -a$, so

$$w(u, a) - u = -\frac{a}{\phi'(\phi^{-1}(\eta))} \tag{3.8}$$

for $\eta$ between $\phi(u) - a$ and $\phi(u)$. As $u \to \bar{u}$, the right side vanishes hence $w(u, a) \to \bar{u}$, and the map $u \mapsto w(u, a)$ is continuous, uniformly for $a$ in a bounded set. Then,

$$\partial_a w(u, a) = -\frac{1}{\phi'(w)}, \qquad \partial_u w(u, a) = \frac{\phi'(u)}{\phi'(w)} > 0, \qquad \text{if } g(u) \neq 0. \tag{3.9}$$

We claim that the map $u \mapsto w(u, a)$ is locally Lipschitz continuous, as well as its inverse, across the singular point $\bar{u}$; that is,

$$\boxed{\forall (u, a) \text{ in a bounded set}, \quad \exists c_1, C_1 \text{ such that } 0 < c_1 \leq \partial_u w(u, a) \leq C_1.}$$

This property is necessary in the proof of the stability result stated in Theorem 3.1 (see e.g. [10, (3.39)]). Indeed, one must check that $\nabla w$ is bounded in a neighborhood of $(\bar{u}, a)$ with $g(\bar{u}) = 0$: if $\partial_a w(u, a) \to 0$ simply as $u \to \bar{u}$, about $\partial_u w$, as in (3.8),

$$w(u, a) - u = -\frac{a}{\phi'(\xi)} = -a\frac{g(\xi)}{f'(\xi)} = -a\frac{g(\xi) - g(\bar{u})}{f'(\xi)},$$

for some $\xi$ intermediate between $w$ and $u$, and then

$$|w - u| = |a|\frac{|g'(\eta)|}{f'(\xi)}|\xi - \bar{u}| \leq |a|\frac{|g'(\eta)|}{f'(\xi)}\left(|w - u| + |u - \bar{u}|\right)$$

for some $\eta = \eta(u, a) \to \bar{u}$ as $u \to \bar{u}$. Hence, if $g'(0) = 0$, we get

$$|w - u|\left(1 - |a|\frac{|g'(\eta)|}{f'(\xi)}\right) \leq |a|\frac{|g'(\eta)|}{f'(\xi)}|u - \bar{u}|$$

leading to

$$\frac{w(u, a) - \bar{u}}{u - \bar{u}} = 1 + \frac{w(u, a) - u}{u - \bar{u}} \to 1 \qquad \text{as } u \to \bar{u}.$$

If $g'(\bar{u}) \neq 0$, it is not true in general that $\partial_u w(\bar{u}, a) = 1$, so it suits to rewrite (3.4) as

$$\int_u^{w(u,a)} \frac{f'(s)}{g(s)}\, ds = -a. \tag{3.10}$$

Since

$$\frac{f'(s)}{g(s)} = \frac{f'(\bar{u})}{g'(\bar{u})(s - u)} + \mathcal{O}(1) \qquad \text{as } u \to \bar{u},$$

then, using (3.10) and that $(w - \bar{u})(u - \bar{u}) > 0$, we deduce that

$$\frac{f'(\bar{u})}{g'(\bar{u})} \log\left(\frac{w - \bar{u}}{u - \bar{u}}\right) + \mathcal{O}(1)(w - u) = -a.$$

Therefore

$$\frac{w - \bar{u}}{u - \bar{u}} = \mathrm{e}^{-a\frac{g'(\bar{u})}{f'(\bar{u})} + \mathcal{O}(1)(w-u)} \to \mathrm{e}^{-a\frac{g'(\bar{u})}{f'(\bar{u})}} > 0 \qquad \text{as } u \to \bar{u}.$$

This proves that $\partial_u w$ exists at $(\bar{u}, a)$ and that it is positive.

### 3.1.2 Wave-Front Tracking Approximations

The wave-front tracking technique provides a standard tool for the analysis of solutions to conservation laws. It was introduced in [6] for scalar conservation laws without source terms and later extended to the case of systems (see the reference books [5, 11, 12]) and to equations/systems with coefficients depending on $x$ (see for instance [13, 15]). Hereafter a wave-front tracking algorithm is described for the Eq. (3.1), as defined in [10, Sect. 2.1]. The basic steps are the following:

- **Step 1**. We fix a partition of $[\bar{w}_1, \bar{w}_2]$: $\mathscr{P} = \{\bar{w}_1 = w_0, \dots, \bar{w}_2 = w_n\}$ and let $\delta > 0$ stand for the corresponding mesh parameter:

$$\delta \doteq \max\{w_i - w_{i-1}\}. \tag{3.11}$$

  This will be used to approximate rarefaction fans. Notice that the partition on $w$ induces a partition on the $u$ axis that depends on $a$: indeed, using $w_u > 0$, define

$$u = P(w; a)$$

  as the inverse function of $u \mapsto w(u, a)$ so that the partition for the $u$ variable is

$$\widetilde{\mathscr{P}}(a) = \{P(w_0; a), \dots, P(w_n; a)\}.$$

- **Step 2**. We define a piecewise constant Riemann solver. Given $U_\ell = (a_\ell, u_\ell)$ and $U_r = (a_r, u_r)$, we consider the usual Riemann problem

$$(a, u)(0, x) = U_\ell \quad \text{for } x < 0, \qquad (a, u)(0, x) = U_r \quad \text{for } x > 0.$$

  If $w_\ell = w(a_\ell, u_\ell)$ and $w_r = w(a_r, u_r) \in \mathscr{P}$, then the piecewise constant Riemann solver still takes values in $\mathscr{P}$. The solution is made of:

  (i) A single steady wave connecting $(a_\ell, w_\ell)$ to $(a_r, w_\ell)$.
  (ii) One or more waves connecting $(a_r, w_\ell)$ to $(a_r, w_r)$. To do this, let $\widetilde{\mathscr{P}}(a_r) = \{u_0, \dots, u_n\}$ the partition on $u$ corresponding to $a_r$ and let $\bar{f}$ be the linear interpolation of $f$ such that $\bar{f}(u_j) = f(u_j)$.

  Then, for $x > 0$, the approximate solution $u$ is the *exact* solution of

$$\partial_t u + \partial_x \bar{f}(u) = 0, \qquad u(0, x) = \begin{cases} P(w_\ell, a_r) & \text{for } x < 0, \\ u_r & \text{for } x > 0. \end{cases}$$

  Such a solution is piecewise constant, valued in $\mathscr{P}$ with waves of positive speed.
- **Step 3**. We consider a piecewise constant initial data $a(x)$, $u_0(x)$ such that $w_0(x) \doteq w(a, u_0)(x)$ is valued in the partition $\mathscr{P}$. At each point of discontinuity of $(a(x), w_0(x))$, the corresponding Riemann problem is solved. The solution

is then defined up to the first time at which an interaction among existing waves occurs; to extend the solution beyond that time, consider new Riemann problems arising at interaction points and solve them according to the method formerly described.

Following [10, Lemma 2.1], interactions among wave fronts are finite and so this approximate solution is defined for all $t \geq 0$ with values in $\mathscr{P}$. The total variation of the Riemann invariant $w(t,\cdot)$ is non-increasing in time. The mesh parameter $\delta > 0$ concerns only the $z$, and so $u$, variables, while $a(x)$ is only piecewise constant.

### 3.1.3 Stability Estimates, Accretive Case

Let $a(x)$, $b(x)$ be piecewise constant, and two wave-front tracking approximations

$$U_1(t,x) = (b,v)(t,x), \qquad U_2(t,x) = (a,u)(t,x),$$

be as defined in Sect. 3.1.2. Let $z(t,x)$ be the Riemann coordinate (see (3.4)) related to $U_1 = (b,v)$. Following [10], we introduce the weight functions

$$W_1(t,x) = \kappa_1 \sum_{y<x} |\Delta z(t,y)| \tag{3.12}$$

$$W_2(x) = \exp\left(\kappa_2 \sum_{y>x} |\Delta a(y)|\right) \tag{3.13}$$

where $\kappa_1$, $\kappa_2$ are constant values to be determined. Then we define the functional

$$\boxed{\Lambda(t;U_1,U_2) = \int_{x_1+Lt}^{x_2} W_1(t,x)W_2(x)|p(x)| + W_2(x)|q(t,x)|\,dx,} \tag{3.14}$$

where $x_1 < x_2$, time $t$ is such that $x_1 + Lt < x_2$,

$$L = \max_{(a,w(u,a))\in K} f'(u) \tag{3.15}$$

and

$$p(x) = a(x) - b(x), \qquad q(t,x) = u(t,x) - \tilde{v}(t,x) \tag{3.16}$$

with

$$\tilde{v}(t,x) = \varphi(a(x); b(x), v(t,x)). \tag{3.17}$$

Here above, $\varphi$ represents the trajectory of the ODE (3.6) issued at $(b(x), v(t,x))$. Another way to express $\tilde{v}$ is by means of the Riemann invariant, see (3.4), and reads:

$$\tilde{v}(t,x) = \phi^{-1}\left(\phi(v(t,x)) + a(x) - b(x)\right), \qquad \phi' = \frac{f'}{g}. \tag{3.18}$$

The identity here above just asserts that $(b(x), v(t,x))$ and $(a(x), \tilde{v}(t,x))$ belong to the same stationary curve. Results stated in [10, Theorem 3.1 and Corollary 3.4] provide an estimate on the $L^1$ norm of $U_1 - U_2$ in the domain of determinacy. Given $x_1 < x_2$,

$$\left\{(t,x): \quad 0 \le t \le \frac{x_2 - x_1}{L},\ x_1 + Lt < x < x_2\right\}.$$

**Theorem 3.1** (Possibly accretive case, $N > 0$) *Let $U_1 = (b, v)$ and $U_2 = (a, u)$ be two wave-front tracking approximations valued in an invariant domain for (3.3), and $\mathscr{P}_1$, $\mathscr{P}_2$ be their corresponding partitions with $\delta_1, \delta_2 > 0$ the mesh parameters, respectively. Denoting*

$$\rho = \text{TV}\{a; [x_1, x_2]\} = \|k\|_{L^1(x_1,x_2)}, \tag{3.19}$$

$$r_1 = \text{TV}\{z[U_1](0,\cdot); [x_1,x_2]\}, \qquad r_2 = \text{TV}\{z[U_2](0,\cdot); [x_1,x_2]\}, \tag{3.20}$$

*there exists a constant $C > 0$ and a choice of $\kappa_1$, $\kappa_2$ such that the functional (3.14) $\Lambda(t) := \Lambda(t; U_1, U_2)$ satisfies for all $0 \le s \le t \le (x_2 - x_1)/L$:*

$$\frac{\Lambda(t) - \Lambda(s)}{t - s} \le C \cdot \mathrm{e}^{\kappa_2 \rho} \cdot [\delta_1 r_1 + \delta_2 r_2]. \tag{3.21}$$

**Corollary 3.1** *In the assumptions of Theorem 3.1, denote by $\mathscr{I}(t)$ the integral*

$$\mathscr{I}(t) = \int_{x_1+Lt}^{x_2} |u(t,x) - v(t,x)|\, dx, \qquad 0 \le t \le \frac{x_2 - x_1}{L}. \tag{3.22}$$

*Then, (3.21) furnishes a time-linear growth, for suitable constants $C$ and $C'$,*

$$\mathscr{I}(t)\mathrm{e}^{-\kappa_2\rho} - \mathscr{I}(0) \le C'(1 + \kappa_1 r_1) \int_{x_1}^{x_2} |a(x) - b(x)|\, dx + C[\delta_1 r_1 + \delta_2 r_2]t. \tag{3.23}$$

*Proof* First we establish the relation between $\|U_1(t) - U_2(t)\|_{L^1((x_1+Lt,x_2))}$ and $\Lambda(t; U_1, U_2)$. We start by observing that the weight functions are uniformly bounded:

$$0 \le W_1 \le \kappa_1 \text{TV}\{z[U_1](0,\cdot); [x_1,x_2]\} = \kappa_1 r_1, \tag{3.24}$$

$$1 \le W_2 \le \exp\left(\kappa_2 \text{TV}\{a; [x_1,x_2]\}\right) = \mathrm{e}^{\kappa_2\rho}. \tag{3.25}$$

Moreover, recalling (3.16) and (3.18), one has $\phi(\tilde{v}) = \phi(v) + p(x)$ and hence

$$|\tilde{v} - v| \le \frac{1}{\inf|\phi'|}|p| = M|p|$$

where $M$ is defined by

$$\boxed{M \doteq \sup_{u\colon (a,w(u,a))\in K} \frac{|g(u)|}{f'(u)}.} \tag{3.26}$$

Therefore it comes

$$|u-v| \le |q| + M|p|, \qquad |q| \le |u-v| + M|p|. \tag{3.27}$$

From (3.21) one has $\Lambda(t) \le \Lambda(0) + C\,e^{\kappa_2\rho}\cdot[\delta_1 r_1 + \delta_2 r_2]\,t$, so (3.24)–(3.27) lead to

$$\begin{aligned}\mathscr{I}(t) &\le \int_{x_1+Lt}^{x_2} |q|\,dx + M\int_{x_1+Lt}^{x_2} |p|\,dx \\ &\le \Lambda(t) + M\int_{x_1+Lt}^{x_2} |p|\,dx \\ &\le \Lambda(0) + M\int_{x_1+Lt}^{x_2} |p|\,dx + C\cdot e^{\kappa_2\rho}\cdot[\delta_1 r_1 + \delta_2 r_2]\cdot t.\end{aligned} \tag{3.28}$$

On the other hand, again by means of (3.24)–(3.27), one has that

$$\begin{aligned}\Lambda(0) &\le e^{\kappa_2\rho}\left\{\int_{x_1}^{x_2} |q(0,x)|\,dx + \kappa_1 r_1\int_{x_1}^{x_2} |p(x)|\,dx\right\} \\ &\le e^{\kappa_2\rho}\left\{\mathscr{I}(0) + (M+\kappa_1 r_1)\int_{x_1}^{x_2} |p(x)|\,dx\right\}.\end{aligned} \tag{3.29}$$

Therefore, using (3.29) within (3.28), brings the inequality,

$$\mathscr{I}(t) \le e^{\kappa_2\rho}\left\{\mathscr{I}(0) + (2M+\kappa_1 r_1)\int_{x_1}^{x_2} |p(x)|\,dx + C[\delta_1 r_1 + \delta_2 r_2]\,t\right\}$$

that leads to (3.23).

*Remark 3.1* The $L^1$ estimate (3.23) displays a substantial improvement with respect to stability results relying on Kružkov's theory, as for instance [14, Theorem 1.2], at least in terms of dependence on $a-b$. Indeed, Kružkov's approach [16] yields an error term proportional to $t\cdot \mathrm{TV}\{a-b\}$ instead of the weaker $\|a-b\|_{L^1}$ in (3.23).

### 3.1.4 Stability Estimates, Non Accretive Case

In this subsection we focus on the case of a non-accretive source term, that is

$$N = \sup_{x,u} k(x)g'(u) \le 0 \qquad \forall x,\ u. \tag{3.30}$$

The functional (3.14) simplifies by selecting $W_2 = 1$, that is $\kappa_2 = 0$, instead of (3.13),

$$\boxed{\Lambda(t; U_1, U_2) = \int_{x_1+Lt}^{x_2} W_1(t,x)|p(x)| + |q(t,x)|\,dx.} \tag{3.31}$$

As a consequence, a **more accurate estimate** can be devised: it is completely analogous to (3.21) and (3.23), with the difference that the multiplying term $e^{\kappa_2 \rho}$ is replaced by 1. The result corresponding to Corollary 3.1 now follows.

**Corollary 3.2** *Under the hypotheses of Theorem 3.1 and assuming (3.30), the following estimate holds for $\mathscr{I}(t)$ given in (3.22):*

$$\boxed{\mathscr{I}(t) \le \mathscr{I}(0) + C'(1+\kappa_1 r_1) \int_{x_1}^{x_2} |a(x) - b(x)|\,dx \; + \; C \cdot [\delta_1 r_1 + \delta_2 r_2] \cdot t} \tag{3.32}$$

*for a suitable constant $C'$, and for all $t$: $0 \le t \le (x_2 - x_1)/L$.*

All the quantities $\kappa_1, \delta_i, r_i, C, C'$ in (3.32) *do not depend on time.*

For convenience, a formal derivation of the inequality (3.32) is yet provided, focusing on the special case of source term being only space-dependent:

$$\partial_t u + \partial_x f(u) = k(x). \tag{3.33}$$

Assumptions in (3.2) are kept, along with a simpler Riemann invariant, $w(u,a) = f(u) - a$. For smooth solutions, the Lyapunov functional $\Lambda$ in (3.31) is shown to be non-increasing, provided $\kappa_1$ is large enough. Recall (3.15)–(3.17) and let $z(t,x)$ be the Riemann coordinate related to $U_1 = (b, v)$:

$$z(t,x) = f(v(t,x)) - b(x).$$

The argument decomposes into several steps.

1. First, we are going to obtain an equation satisfied by $\tilde{v}$, defined by

$$\tilde{v} = f^{-1}(f(v) + a - b) = f^{-1}(z + a). \tag{3.34}$$

Since $z$ is constant along $v$-characteristics: $\left(\partial_t + f'(v)\partial_x\right) z = 0$, we find that

$$\begin{aligned}\left(\partial_t + f'(\tilde{v})\partial_x\right) f(\tilde{v}) &= \left(\partial_t + f'(\tilde{v})\partial_x\right) z + f'(\tilde{v})\partial_x a \\ &= \left(f'(\tilde{v}) - f'(v)\right)\partial_x z + f'(\tilde{v})\partial_x a\end{aligned}$$

and therefore

$$\left(\partial_t + f'(\tilde{v})\partial_x\right)\tilde{v} = \frac{f'(\tilde{v}) - f'(v)}{f'(\tilde{v})}\partial_x z + \partial_x a.$$

2. We claim that

$$\partial_t|u - \tilde{v}| + \partial_x|f(u) - f(\tilde{v})| \le C|a - b|\,|\partial_x z| \tag{3.35}$$

for a suitable constant $C > 0$. Indeed, consider the equations for $u$ and $\tilde{v}$:

$$\begin{aligned} \partial_t u + \partial_x f(u) &= \partial_x a, \\ \partial_t \tilde{v} + \partial_x f(\tilde{v}) &= \partial_x a + \frac{f'(\tilde{v}) - f'(v)}{f'(\tilde{v})}\partial_x z. \end{aligned}$$

Notice that the term $\partial_x a$ cancels in the difference of the above equations. To evaluate $\partial_t|u - \tilde{v}|$, use that $f' > 0$ and get

$$\partial_t|u - \tilde{v}| + \partial_x|f(u) - f(\tilde{v})| \le \frac{\left|f'(\tilde{v}) - f'(v)\right|}{f'(\tilde{v})}\,|\partial_x z|\,. \tag{3.36}$$

About the last term in (3.36), by using (3.34) one obtains

$$|a - b| = |f(\tilde{v}) - f(v)| \ge \inf f' \cdot |\tilde{v} - v| \;=\; c\,|\tilde{v} - v|$$

with

$$c = \min\{f'(u);\ u : (a, w(u, a)) \in K \text{ for some } a\}.$$

Notice that $c > 0$, since $f'$ is bounded away from zero. Hence we deduce

$$\frac{|f'(\tilde{v}) - f'(v)|}{f'(\tilde{v})} \le \frac{\sup|f''|}{c}\,|\tilde{v} - v| \;\le\; C|a - b|$$

with $C = \sup|f''|/c^2 > 0$. This completes the proof of (3.35).

3. We recall the definition of the weight $W_1$, see (3.12):

$$W_1(t, x) = \kappa_1 \mathrm{TV}\{z(t, \cdot); (x_1 + Lt, x)\}.$$

It will be used to balance the term appearing in (3.35). We will use that

$$\partial_t W_1(t, x) \;\le\; -\kappa_1 c\,|\partial_x z|. \tag{3.37}$$

Indeed, recalling that $z$ is constant along $v$-characteristics, deriving by $x$ the equation $z_t + f'(v)z_x = 0$ and by setting $q = z_x$, we find

$$\partial_t q + \partial_x[f'(v)q] = 0.$$

Multiplying by $\mathrm{sgn}(q)$ gives $|q|_t + \left[f'(v)|q|\right]_x = 0$. We can now evaluate $\partial_t W_1$:

$$\begin{aligned}
\frac{1}{\kappa_1}\partial_t W_1(t,x) &= \partial_t \left\{ \int_{x_1+Lt}^{x} |z_x(t,y)|\,dy \right\} \\
&= \int_{x_1+Lt}^{x} \partial_t |z_x(t,y)|\,dy - L|z_x(t,x_1+Lt)| \\
&= -\int_{x_1+Lt}^{x} \partial_x \left[f'(v)|z_x(t,y)|\right] dy \;-\; L|z_x(t,x_1+Lt)| \\
&= -f'(v(t,x))|z_x(t,x)| + \underbrace{\left(f'(v(t,x_1+Lt)) - L\right)}_{\le 0} |z_x(t,x_1+Lt)| \\
&\le -f'(v(t,x))|z_x(t,x)| \;\le\; -c|z_x(t,x)|
\end{aligned}$$

by definition (3.15) of $L$. Hence (3.37) is proved.

4. Now, combining (3.35) and (3.37), we get that

$$\partial_t\{|u-\tilde{v}| + W_1|a-b|\} + \partial_x\{|f(u) - f(\tilde{v})|\} \le 0. \tag{3.38}$$

Indeed

$$\begin{aligned}
&\partial_t\{|u-\tilde{v}| + W_1|a-b|\} + \partial_x\{|f(u)-f(\tilde{v})|\} \\
&\quad = \partial_t\{|u-\tilde{v}|\} + \partial_x\{|f(u)-f(\tilde{v})|\} + \partial_t\{W\}|a-b| \\
&\quad \le C|a-b||\partial_x z| - \kappa_1 c|a-b||\partial_x z| \le 0
\end{aligned}$$

for $\kappa_1 \ge C/c = (\sup|f''|)/c^3$.

Now we are ready to study the evolution in time for $\Lambda$, (3.31), and check the contribution of the boundary terms. Using (3.38) and definition of $L$ in (3.15),

$$\begin{aligned}
\partial_t \Lambda(t) &\le -|f(u) - f(\tilde{v})||_{x=x_1+Lt}^{x=x_2} \;-\; L\{|u-\tilde{v}| + W_1|a-b|\}|_{x=x_1+Lt} \\
&\le 0 + \{|f(u)-f(\tilde{v})| \;-\; L|u-\tilde{v}|\}\,|_{x=x_1+Lt} \\
&\le 0.
\end{aligned}$$

This concludes the formal proof.

### 3.1.5 Limit $\delta \to 0$ and Recovery of Kružkov's Entropy Solution

Let $U_1$, $U_2$ be wave-front tracking approximations of the same exact solution, associated to the data $(a_0, u_0)$. By letting $\delta_2 \to 0$ in either (3.23) or (3.32), $U_2$ approaches the exact, entropy solution $(a_0(x), u(t,x))$. Thus an error estimate for the wave-front tracking scheme easily follows, see next Corollary 3.3. In order to achieve convergence, we need to specify how the initial data are approximated. Set $\delta = \delta_1$ and

$$b(x) = a_0(j\delta), \quad v(0,x) = u_0(j\delta) \qquad \text{for } x \in [j\delta, (j+1)\delta).$$

Let $\mathscr{P}$ be any partition of $[\bar{w}_1, \bar{w}_2]$ with mesh parameter $\leq \delta$ that includes all the points $w\,(u_0(j\delta), a_0(j\delta))$, as $j$ varies in $\mathbb{Z}$. We state the result related to (3.23); a completely analogous result would be obtained starting from (3.32).

**Corollary 3.3** *Let $u(t,x)$ be a solution to (3.1) and $a_0(x) = \int_{-\infty}^{x} k(s)\,ds$. Denote $w_0 = w(u_0, a_0)$ and let $(v, b)$ be the approximate solution with parameter $\delta > 0$ corresponding to the data above. Let $x_1 \in \delta\mathbb{Z}$, $x_1 < x_2$, $Lt \leq x_2 - x_1$. Then the following inequality holds, for suitable constants $C$ and $C'$:*

$$e^{-\kappa_2 \rho} \int_{x_1+Lt}^{x_2} |u(t,x) - v(t,x)|\,dx \leq \delta\, C\, r \cdot t + \delta\left[\mathrm{TV}\{u_0; [x_1, x_2]\} + C'(1+\kappa_1 r)\rho\right], \tag{3.39}$$

*where* $\rho = \mathrm{TV}\{a_0; [x_1, x_2]\} = \|k\|_{L^1(x_1,x_2)}$, $r = \mathrm{TV}\{w_0; [x_1, x_2]\}$.

*Proof* We first consider a sequence $\delta_{2,k} \to 0$ and perform the limit in (3.23), similarly as done in [10, Theorem 4.1]. Given a sequence of partitions $\mathscr{P}_k$, with corresponding $\delta_{2,k} \to 0$ as $k \to \infty$, we choose $a_{0,k}$, $u_{0,k}$ piecewise constant and such that:

$$a_{0,k} \to a_0, \quad u_{0,k} \to u_0 \quad \text{in } L^1_{loc}, \qquad w_{0,k} = w(a_{0,k}, u_{0,k}) \in \mathscr{P}_k,$$

and, for some $R$ independent on $k$:

$$\begin{aligned} &\mathrm{TV}\{a_{0,k}\} \leq \mathrm{TV}\{a_0\}, \qquad \mathrm{TV}(a_{0,k}, u_{0,k}) \leq R, \\ &\limsup_{k\to\infty} \mathrm{TV} w(a_{0,k}, u_{0,k}) \leq \mathrm{TV} w(a_0, u_0). \end{aligned}$$

Choosing the approximation of $(a_0, u_0)$ as above, $\rho_k \leq \rho$ and $r_{2,k}$ is uniformly bounded (see (3.19) and (3.20)). Moreover, the constant values $C, C'$ are uniform in $k$. Therefore, passing to the limit in (3.23),

$$\begin{aligned} &e^{-\kappa_2 \rho} \int_{x_1+Lt}^{x_2} |u(t,x) - v(t,x)|\,dx \\ &\leq \int_{x_1}^{x_2} |u_0(x) - v(0,x)|\,dx + C'(1+\kappa_1 r) \int_{x_1}^{x_2} |a_0(x) - b(x)|\,dx + C\,\delta\, r\, t. \end{aligned} \tag{3.40}$$

Yet, using that $x_1 \in \delta\mathbb{Z}$, notice that

$$\int_{x_1}^{x_2} |u(0,x) - v(0,x)|\, dx \le \delta\, \{\mathrm{TV}\{u_0; [x_1, x_2)\}\},$$

$$\int_{x_1}^{x_2} |a_0(x) - b(x)|\, dx \le \delta\, \{\mathrm{TV}\{a_0; [x_1, x_2)\}\},$$

and substitute in (3.40) in order to obtain, as $\delta \to 0$:

$$\mathrm{e}^{-\kappa_2 \rho} \int_{x_1+Lt}^{x_2} |u(t,x) - v(t,x)|\, dx \le \delta \left[\mathrm{TV}\{u_0; [x_1, x_2)\} + C'(1+\kappa_1 r)\rho + C\, r\, t\right]$$

that is, (3.39).

## 3.2 Periodic Version of Wavefront Tracking

### *3.2.1 A 1-Periodic, Non-Accretive Equation*

Let's switch onto a special case of the Cauchy problem for (3.33), rewritten as

$$\partial_t u + \partial_x \{f(u) - a(x)\} = 0,$$

with

$$f' > 0, \qquad f^{-1} : \mathbb{R} \to \mathbb{R}, \quad a(x) \in C^1(\mathbb{R}/\mathbb{Z}), \quad \int_0^1 a(x)\, dx = 0. \tag{3.41}$$

Clearly also $\int_0^1 a'(x)\, dx = 0$ thanks to the 1-periodicity of $a$. Assume moreover that the initial data $u_0$ is 1-periodic, then, the corresponding Cauchy problem admits a unique admissible (entropy) solution which is 1-periodic in space. Riemann coordinates for the $2 \times 2$ system,

$$\partial_t u + \partial_x \{f(u) - a\} = 0, \qquad \partial_t a = 0 \tag{3.42}$$

can be chosen as $a$, $w(u,a) = f(u) - a$. The transformation $(a,u) \mapsto (a,w)$ is a diffeomorphism from $\mathbb{R}^2$ to $\mathbb{R}^2$. Since $|f| \to \infty$ as $|u| \to \infty$, then it maps bounded sets into bounded sets.

*Remark 3.2* Under the same monotonicity condition (3.41) for $f$, the Cauchy problem for (3.42) was solved in [3], where authors prove the existence of solutions with data $u_0$ and forcing $a(x)$ both in $L^1 \cap L^\infty$, as well as their continuity with respect to perturbations in the $L^1$ norm for both $u_0$ and $a$, [3, Theorem 1.1].

### 3.2.2 A Specific 1-Periodic Lyapunov Functional

Provided that the initial data $u_0$ is 1-periodic, one can define approximate solutions to the Cauchy problem for (3.42) by following the algorithm in Sect. 3.1.2. In particular, by choosing approximate initial data which are 1-periodic, one obtains approximate solutions that keep the same space-periodicity; in addition, the total variation of the Riemann invariant $t \mapsto w(u, a)(t, \cdot)$ will be non-increasing when restricted to a complete period $\mathbb{R}/\mathbb{Z}$. To investigate the stability of these approximations, one may employ the functional $\Lambda$ given in (3.31). However, it does not exploit the property of 1-periodicity in space of the approximate solutions. Indeed, a direct application of (3.32) is not convenient, in order to estimate $\int_0^1 |u(t,x) - v(t,x)|\,dx$: the $L^1$ norms on the right hand side of (3.32), of $(u(0,x) - v(0,x))$ and of $a - b$, would be computed on the interval $(-Lt, 1)$, leading to a linear increase in $t$ (see [17]) with respect to the norm of these quantities on $\mathbb{R}/\mathbb{Z}$. Instead, we define

$$\Lambda_1\left(U_1(t,\cdot), U_2(t,\cdot)\right) = \int_0^1 W_1(t,x)|a(x) - b(x)| + |u(t,x) - \tilde{v}(t,x)|\,dx,$$

$$W_1(t,x) = \kappa_1 \sum_{-1+Lt<y<x} |\Delta z(t,y)| = \kappa_1 \mathrm{TV}\{z(t,\cdot); (-1+Lt, x)\}, \tag{3.43}$$

for $t \le 1/L$. More generally, for $\frac{k}{L} < t < \frac{k+1}{L}$, we set

$$\boxed{W_1(t,x) = \kappa_1 \mathrm{TV}\{z(t,\cdot); (-1+(Lt-k), x)\}, \qquad k < Lt < k+1,} \tag{3.44}$$

where $z$ is the Riemann coordinate related to $U_1$. Following the formal argument described in Sect. 3.1.4, we notice that $W_1$ is defined for all $x \in (0,1)$ whenever $0 < Lt - k < 1$, and its time-derivative satisfies the estimate (3.37), since it is simply the special case of $x_1 = -1$. Therefore we can use (3.38) and get, for $k < Lt < k+1$:

$$\begin{aligned}
\partial_t \Lambda_1(t) &= \int_0^1 \partial_t\left\{W_1(t,x)|a(x) - b(x)| + |u(t,x) - \tilde{v}(t,x)|\right\}dx \\
&\le -\int_0^1 \partial_x\left\{|f(u) - f(\tilde{v})|\right\}dx \\
&= -|f(u) - f(\tilde{v})|\big|_{x=0}^{x=1} \\
&= 0.
\end{aligned}$$

More precisely, a slight variant of the proof of Theorem 3.1 (which is given in [10]), allows us to prove an analogue of (3.21): in the notation of Theorem 3.1, one has

$$\frac{\Lambda_1(t) - \Lambda_1(s)}{t-s} \le C \cdot [\delta_1 r_1 + \delta_2 r_2] \qquad k/L \le s \le t \le (k+1)/L$$

where $\delta_1$, $\delta_2 > 0$ are the mesh parameters corresponding to the partitions of the wave-front tracking approximation, while

$$r_1 = \mathrm{TV}\{w[U_1](t = k/L+, \cdot); \mathbb{R}/\mathbb{Z}\}, \qquad r_2 = \mathrm{TV}\{w[U_2](t = k/L+, \cdot); \mathbb{R}/\mathbb{Z}\},$$

which are bounded by the corresponding quantities at time $t = 0$:

$$r_1 \le r_{1,0} \doteq \mathrm{TV}\{w[U_1](0, \cdot); \mathbb{R}/\mathbb{Z}\}, \qquad r_2 \le r_{2,0} \doteq \mathrm{TV}\{w[U_2](0, \cdot); \mathbb{R}/\mathbb{Z}\}.$$

In general $\Lambda_1(t)$ will be discontinuous at each time $t > 0$: $Lt \in \mathbb{Z}$. Then, we will need to evaluate how the weight $W_1$ changes at time $t = k/L$:

$$\begin{aligned}
&\frac{1}{\kappa_1}\left(W_1(k/L+, x) - W_1(k/L-, x)\right) \\
&\quad = \mathrm{TV}\{z(k/L+, \cdot); (-1, 0]\} + \mathrm{TV}\{z(k/L+, \cdot); (0, x)\} - \mathrm{TV}\{z(k/L-, \cdot); [0, x)\} \\
&\quad \le \mathrm{TV}\{z(k/L-, \cdot); (-1, 0]\} + \underbrace{\mathrm{TV}\{z(k/L+, \cdot); (0, x)\} - \mathrm{TV}\{z(k/L-, \cdot); [0, x)\}}_{\le 0} \\
&\quad \le \mathrm{TV}\{z(0, \cdot); (-1, 0]\} = r_{1,0},
\end{aligned}$$

where we used the decreasing properties of the total variation of the Riemann invariant. Therefore, an estimate for $\Lambda_1$ writes as

$$\begin{aligned}
\Lambda_1(t) &\le \Lambda_1(0) + \sum_{1\le k\le N} [\Lambda_1(k/L+) - \Lambda_1(k/L-)] + Ct[\delta_1 r_{1,0} + \delta_2 r_{2,0}] \\
&\le \Lambda_1(0) + \kappa_1 \int_0^1 \sum_{1\le k\le N} r_{1,0}|a(x) - b(x)|\,dx + Ct[\delta_1 r_{1,0} + \delta_2 r_{2,0}] \\
&\le \Lambda_1(0) + t\left\{\kappa_1 r_{1,0} L \int_0^1 |a(x) - b(x)|\,dx + C[\delta_1 r_{1,0} + \delta_2 r_{2,0}]\right\} \qquad (3.45)
\end{aligned}$$

where $N$ is the integer part of $Lt$, that is, $N \in \mathbb{N}$ and $N \le Lt < N + 1$. Moreover

$$\begin{aligned}
\Lambda_1(0) &\le 2\kappa_1 \tilde{r}_1 \int_0^1 |a - b|\,dx + \int_0^1 |u(0, x) - \tilde{v}(0, x)|\,dx \\
&\le (M + 2\kappa_1 \tilde{r}_1) \int_0^1 |a - b|\,dx + \int_0^1 |u(0, x) - v(0, x)|\,dx \qquad (3.46)
\end{aligned}$$

where $M = 1/\inf f' > 0$, see (3.26). In conclusion, with the help of (3.45) and (3.46), one reaches an estimate similar to (3.32). Indeed, for $\mathscr{I}(t) = \int_0^1 |u - v|\,dx$,

$$
\begin{aligned}
\mathscr{I}(t) &\leq \int_0^1 |u-\tilde{v}|\,dx + M\int_0^1 |a-b|\,dx \\
&\leq \Lambda_1(t) + M\int_0^1 |a-b|\,dx \\
&\leq \Lambda_1(0) + (M+\kappa_1\tilde{r}_1 Lt)\int_0^1 |a(x)-b(x)|\,dx + Ct[\delta_1 r_{1,0}+\delta_2 r_{2,0}] \\
&\leq \mathscr{I}(0) + \{2M+\kappa_1\tilde{r}_1(2+Lt)\}\int_0^1 |a-b|\,dx + Ct[\delta_1 r_{1,0}+\delta_2 r_{2,0}].
\end{aligned}
$$

## 3.3 Numerical Validation for Scalar Balance Laws

### *3.3.1 A First Numerical Validation*

Hereafter we aim at studying a benchmark evoked in [13], in a non-resonant framework though, as the numerical assessments in the previous section suggest that discrepancies occur together with nonlinear resonance (loss of strict hyperbolicity), $f'(u)=0$. Let $x\in(-1,1)$ stand for the computational domain with periodic boundary conditions and set up a position-dependent scalar balance law,

$$
\partial_t u + \partial_x f(u) = a'_\beta(x)g(u), \qquad f(u)=4u(1-u), \qquad g(u)=u, \tag{3.47}
$$

and, for $\beta\leq 1$, a negative coefficient $a_\beta - \|a_\beta\|_\infty$ where,

$$
a_\beta(x) = \begin{cases} \frac{3}{2\pi}\cos(\frac{\pi x}{\beta}) & |x|<\beta \\ -\frac{3}{2\pi} & \beta\leq|x|\leq 1. \end{cases}
$$

With such a choice in the functions $f$ and $g$, the expression of the Riemann invariant $w(u,a)=\phi^{-1}(\phi(u)-a)$ is more involved and requires the so-called "Lambert W-function", (see [4] or [9, p. 256]). To be more specific, up to a constant,

$$
\phi'(u) = \frac{f'(u)}{g(u)} = \frac{4(1-2u)}{u}, \qquad \phi(u)-a = 4(\log|u|-2u)-a.
$$

Accordingly, the expression of the Riemann invariant is:

$$
w(u,a) = -\frac{1}{2}W\big(-2\exp(\log|u|-2u-a/4)\big),
$$

$W$ standing for the real-valued branch such that $W(0)=0$. Observe that it may be necessary to further take its real part when inserting it inside a computer program. As long as $f'(u) > 0$, an expression of the $L^1$ error of WB-WFT was given by Corollary 1 in [1, p. 482]. Still, we set up periodic initial data for (3.47) as,

$$u(t = 0, x) = u_0(x) = \frac{1}{4}\big(1 + 0.9\sin(\pi x)\big). \tag{3.48}$$

On Fig. 3.1, we display some numerical results obtained by WB-WFT: convective fronts associated to the genuinely non-linear field are drawn in blue. Zero-waves

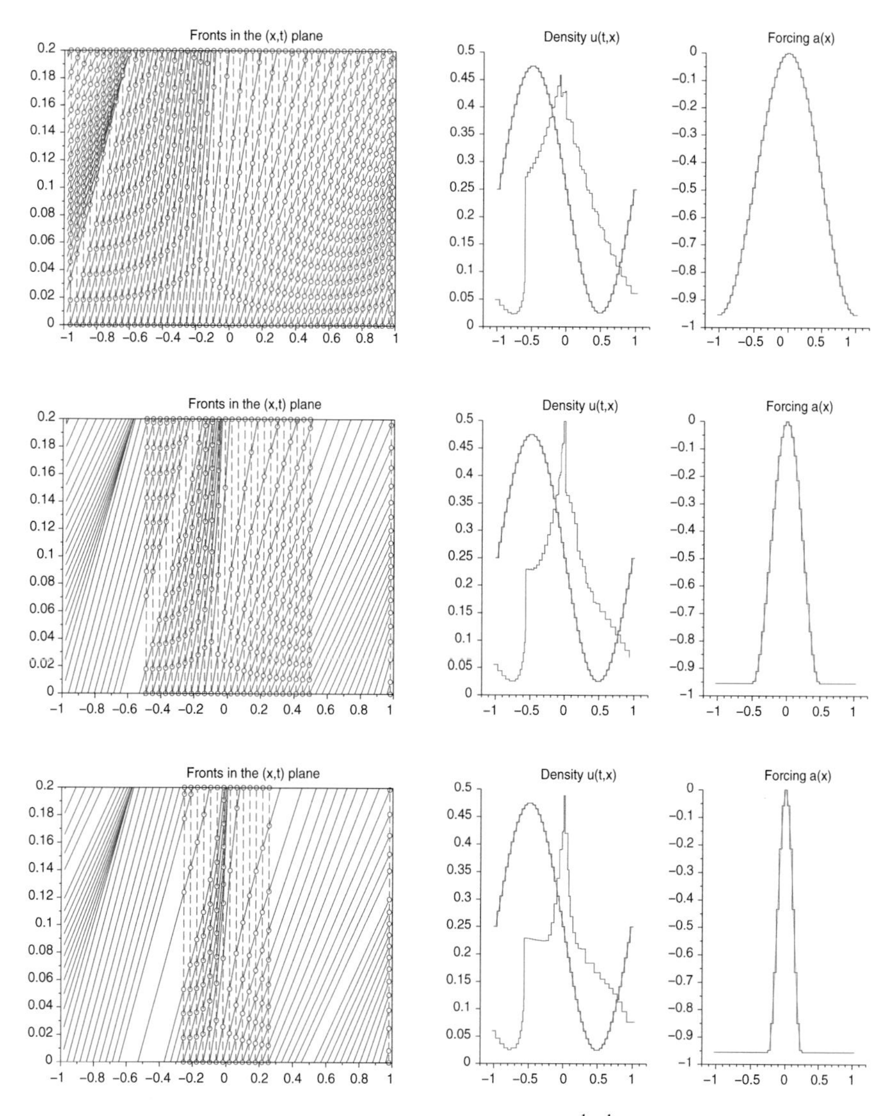


**Fig. 3.1** Outcomes of WB-WFT with (3.47)–(3.48) for $\beta = 1, \frac{1}{2}, \frac{1}{4}$ (From *top* to *bottom*)

correspond to dashed red lines. Each red circle stands for an interaction between a convective front and a standing wave. The test-case's stiffness in space automatically decreases with $\beta$. There, we considered only, $\beta = \frac{1}{4}, \frac{1}{2}, 1$ with $N = 51$ initial fronts locations and $\delta = 0.07$.

### 3.3.2 Numerical Validation for Periodic Forcing

Hereafter, the scalar law (3.1) is recast into the periodicity interval $x \in (-1, 1)$ for convenience. In order to produce an exact solution, to scrutinize the $L^1$ errors and to study the accuracy of our estimates, let us define for any $0 < \beta < 1$:

$$\forall x \in (-1, 1), \qquad 0 \le a_\beta(x) = \begin{cases} \frac{x^2}{2\beta^2} & |x| \le \beta, \\ \frac{1}{2} & \beta < |x| < 1. \end{cases} \tag{3.49}$$

The sequence of non-negative $a_\beta$ is such that

$$\text{TV}a_\beta \equiv 1, \qquad \max(a_\beta) \equiv \frac{1}{2}.$$

We seek the exact solution in $x \in (-1, 1)$ of the Cauchy problem,

$$\partial_t u + \partial_x f(u) = a'_\beta(x), \qquad u_0(x) \equiv 1, \qquad f(u) = \frac{u^2}{2}. \tag{3.50}$$

Inside the region where the forcing term acts, the differential system of characteristics curves $t \mapsto X(t, x_0)$ reads,

$$\dot{X} = U, \qquad \dot{U} = a'_\beta(X) = \alpha X, \qquad U(t, x_0) = u(t, X(t, x_0)) \text{ for } x_0 \in (-1, 1).$$

where $\alpha = \frac{1}{\beta^2}$. By (3.50), for any $x_0 \in (-1, 1)$, we have $U(t = 0, x_0) = 1$. Let the final time $T > 0$ be chosen before the shock onset, a tedious computation yields:

- Consider the times

$$t_\beta^- = \max(0, \min(T, -\beta - x_0)), \quad t_r = \max(0, \min(t_\beta^+, \max(0, T - t_\beta^-))),$$

  where $t_\beta^+$ is the solution of

$$(x_0 + t_\beta^-) \cosh(t\sqrt{\alpha}) + \sinh(t\sqrt{\alpha})/\sqrt{\alpha} = x_0.$$

- The exact value reads:

$$U(T, x_0) = U(t = 0, x_0) \cosh(t_r \sqrt{\alpha}) + (x_0 + t_\beta^-) \sinh(t_r \sqrt{\alpha}) \sqrt{\alpha}.$$

- At the location

$$X(T, x_0) = (x_0 + t_\beta^-) \cosh(t_r \sqrt{\alpha}) + \frac{\sinh(t_r \sqrt{\alpha})}{\sqrt{\alpha}} + U(T, x_0)(T - t_r) \chi(x_0 > -\beta).$$

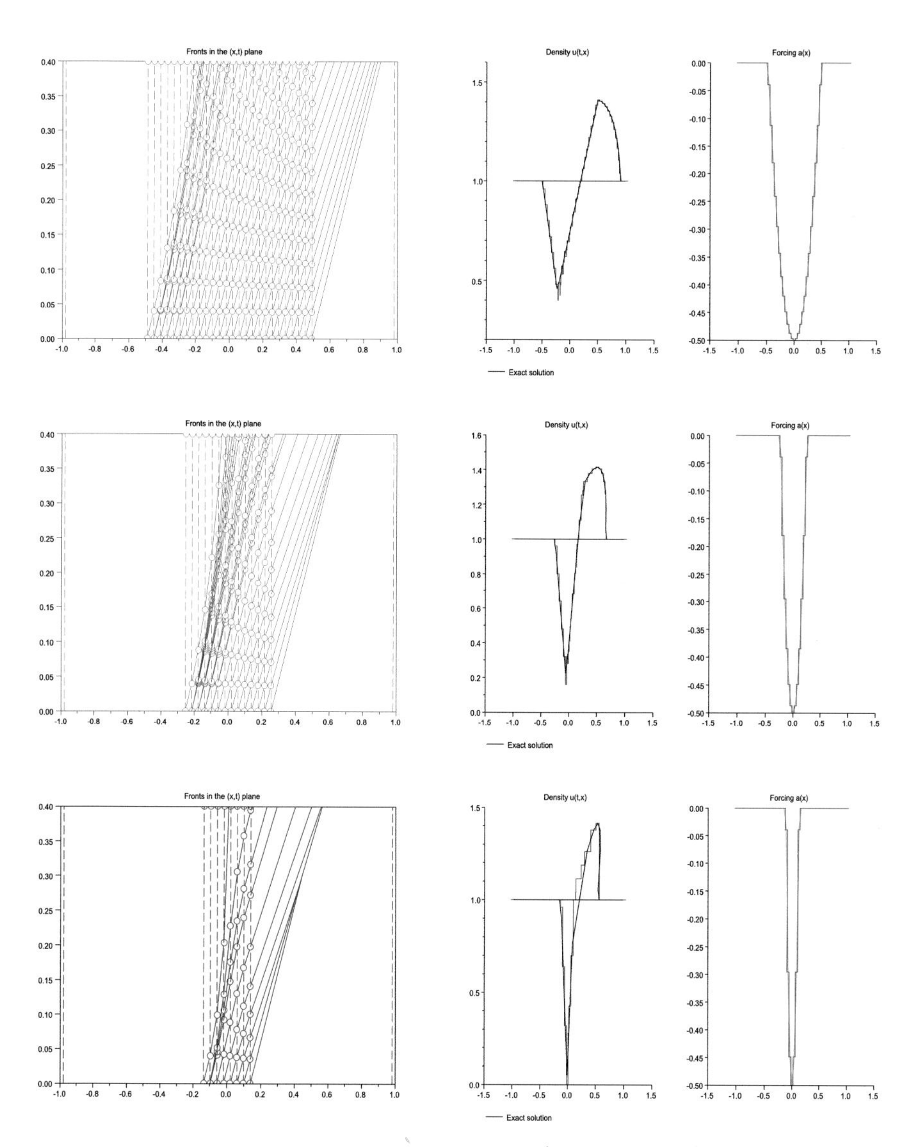


**Fig. 3.2** Fronts (*left*) and approximations (*right*) for $\beta = \frac{1}{2}$ (*top*) and $\beta = \frac{1}{4}$ (*bottom*)

If $X(T, x_0) > 1$, then one simply translates by periodicity $X(T, x_0) - 2$.

We may also consider the case in which the source term is $-a'_\beta(x)$. Correspondingly we have $\alpha = -1/\beta^2 < 0$. The square root is imaginary and hyperbolic trigonometric functions reduce to usual trigonometric ones. The calculation still delivers the corresponding exact solution. The most useful ingredient in this derivation is the analytic expression of the solution to the second-order ODE,

$$X'' = \alpha X, \qquad X(t = 0, x_0) = x_0, \qquad \dot{X}(t = 0, x_0) = U(t = 0, x_0) \equiv 1.$$

### 3.3.3 Practical Measurements of the Global $L^1$ Error

Measuring the global error is a delicate task, mainly because as the wavefront tracking algorithm produces at time $T$ a set of, say $N + 1$ discrete values of $u$ separated at the $N$ front locations, they usually don't meet with the points where characteristics $X(T, \cdot)$ end up. On Fig. 3.2, two values of $\beta$, $\frac{1}{2}$ and $\frac{1}{4}$ were considered, the latter leading to a problem endowed with a greater stiffness in the space variable, meaning that the corresponding Lipschitz constant $\|a'_\beta\|_\infty$ is higher. Pointwise errors are seen on Fig. 3.3: if the left picture shows a quite uniform spreading of the numerical errors when $\beta = \frac{1}{2}$, the right one, in contrast, reveals that errors are peaked around $x = 0$ for $\beta = \frac{1}{4}$, which is a place where the WFT approximation goes below 0.2, so it gets quite close to nonlinear resonance where $f'$ vanishes. Consequently, two types of $L^1$ errors are presented in Table 3.1: the usual one $e_\beta(T)$, and a renormalized one $\tilde{e}_\beta(T) = c_\beta \, e_\beta(T)$ where $c_\beta = \min f'(u(T, \cdot))$, $u$ the exact solution (retrieved for instance by characteristics). The scope is to check whether $\tilde{e}_\beta$ remains roughly constant as $\beta$ varies, thanks to the fact that $\text{TV}a_\beta \equiv 1$ (as suggested by the previous rigorous error estimates). For completeness, we display on Fig. 3.4 the numerical outcome of WB-WFT in case $\alpha < 0$ and characteristics $X(t, x_0)$ are expressed by means of

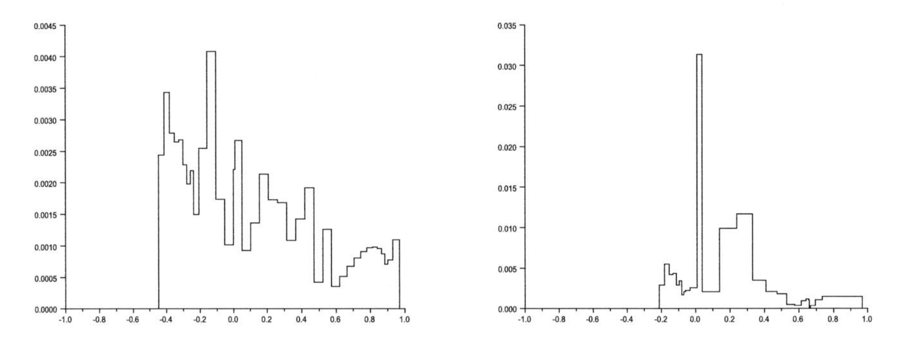


**Fig. 3.3** Pointwise errors for $\beta = \frac{1}{2}$ (*left*) and $\beta = \frac{1}{4}$ (*right*): notice the spike

**Table 3.1** Measured $L^1$ errors $e_\beta(T)$, $\tilde{e}_\beta(T)$ with respect to $\beta$ and $\Delta x$ at $T = 0.4$

| Points | $N = 41$ | | $N = 61$ | | $N = 81$ | |
|---|---|---|---|---|---|---|
| Errors | $e_\beta(T)$ | $\tilde{e}_\beta(T)$ | $e_\beta(T)$ | $\tilde{e}_\beta(T)$ | $e_\beta(T)$ | $\tilde{e}_\beta(T)$ |
| $\beta = \frac{3}{4}$ | 0.070814 | 0.042508 | 0.048064 | 0.028351 | 0.040036 | 0.023764 |
| $\beta = \frac{1}{2}$ | 0.088340 | 0.041607 | 0.054488 | 0.025277 | 0.039917 | 0.018374 |
| $\beta = \frac{1}{4}$ | 0.212905 | 0.050987 | 0.101077 | 0.021436 | 0.067174 | 0.014786 |

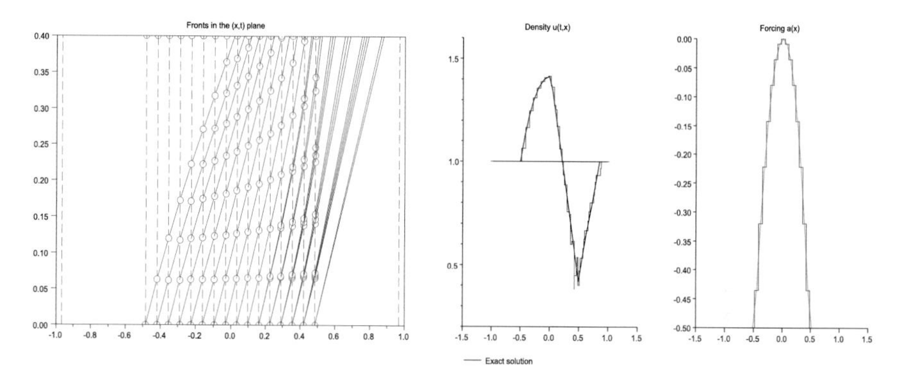


**Fig. 3.4** Numerical results with $\beta = \frac{1}{2}$, but $\alpha < 0$, and $N = 31$, $\delta = 0.02$

usual trigonometric functions. A coarse grid was prescribed (only 31 points initially), together with a small value of $\delta = 0.02$ in order to visualize the multiplication of rarefaction fronts. The comparison with the corresponding exact solution is again satisfying, and very similar to the former case for which $\alpha > 0$.

## References

1. D. Amadori, L. Gosse, Transient $L^1$ error estimates for well-balanced schemes on non-resonant scalar balance laws. J. Differential Equations **255**, 469–502 (2013)
2. D. Amadori, L. Gosse, G. Guerra, Godunov-type approximation for a general resonant balance law with large data. J. Differential Equations **198**, 233–274 (2004)
3. P. Baiti, H.K. Jenssen, Well-posedness for a class of $2 \times 2$ conservation laws with $L^\infty$ data. J. Differential Equations **140**, 161–185 (1997)
4. J.P. Boyd, Global Approximations to the Principal Real-Valued Branch of the Lambert W-function. Applied Math Lett. **11**, 27–31 (1998)
5. A. Bressan, *Hyperbolic Systems of Conservation Laws - The one-dimensional Cauchy problem, Oxford Lecture Series in Mathematics and its Applications 20* (Oxford University Press, Oxford, 2000)
6. C.M. Dafermos, Polygonal approximations of solutions of the initial value problem for a conservation law. J. Math. Anal. Appl. **38**, 33–41 (1972)
7. M. Ersoy, E. Feireisl, E. Zuazua, Sensitivity analysis of 1–d steady forced scalar conservation laws. J. Differential Equations **254**, 3817–3834 (2013)

8. L. Gosse, Localization effects and measure terms in numerical schemes for balance laws. Math. Comp. **71**, 553–582 (2002)
9. L. Gosse, *Computing Qualitatively Correct Approximations of Balance Laws*, vol. 2. (SIMAI Springer Series, Springer, 2013)
10. G. Guerra, Well-posedness for a scalar conservation law with singular nonconservative source. J. Differential Equations **206**, 438–469 (2004)
11. H. Holden, K.H. Karlsen, K.-A. Lie, N.H. Risebro, *Splitting Methods for Partial Differential Equations with Rough Solutions* (European Mathematical Society (EMS), Zürich, Analysis and MATLAB programs. EMS Series of Lectures in Mathematics, 2010)
12. H. Holden, N.H. Risebro, *Front Tracking for Hyperbolic Conservation Laws. Applied Mathematical Sciences*, vol. 152. (Springer-Verlag, New York, 2002)
13. K.H. Karlsen, N.H. Risebro, J.D. Towers, Front tracking for scalar balance equations. J. Hyperbolic Differ. Equ. **1**, 115–148 (2004)
14. R.A. Klausen, N.H. Risebro, Stability of conservation laws with discontinuous coefficients. J. Differential Equations **157**, 41–60 (1999)
15. C. Klingenberg, N.H. Risebro, Convex conservation laws with discontinuous coefficients. Existence, uniqueness and asymptotic behavior. Comm. Partial Differential Equations **20**(11–12), 1959–1990 (1995)
16. S.N. Kružkov, First order quasilinear equations in several independent space variables. Mat. USSR Sbornik **81**, 228–255 (1970)
17. W.J. Layton, Error Estimates for Finite Difference Approximations to Hyperbolic Equations for Large Time. Proc. Amer. Math. Soc. **92**, 425–431 (1984)
18. R. Natalini, A. Tesei, Blow-up of solutions for a class of balance laws. Comm. Part. Diff. Eqns. **19**, 417–453 (1994)
19. E. Weinan, Homogenization of scalar conservation laws with oscillatory forcing terms. SIAM J. Appl. Math. **52**, 959–972 (1992)

# Chapter 4
# Lyapunov Functional for Inertial Approximations

**Abstract** In this chapter we address a semilinear system of two equations, in one space dimension, related to the wave equation with space-dependent damping. An approximation scheme is defined, of Well-Balanced type; for this scheme an error estimate is devised by means of the stability analysis for hyperbolic systems.

**Keywords** Well-balanced schemes · Stability theory applied to numerical schemes · Weakly dissipative hyperbolic systems

The error estimates obtained so far on scalar laws are quite strong: roughly speaking, they express the fact that, even in presence of both an accretive source term $g(u)$ and a position-dependent (but integrable) coefficient $k(x)$, the global $L^1$ error of the numerical process depends neither on an exponential in time (no Gronwall lemma involved) nor on the derivatives of $k$ (just on $\|k\|_{L^1}$), all this for initial data of arbitrary big total variation. The only stringent restriction is the non-resonance assumption, which has a meaningful physical interpretation: waves cannot be trapped inside an area where the source term makes them grow beyond control. Since the characteristic velocity $f'(u)$ never vanishes and $k \in L^1(\mathbb{R})$, convective waves exit from those strongly accretive areas in finite time thus overall stability is kept. A challenging question is therefore: how much of these results carry onto the far more delicate case of 1D, position-dependent, hyperbolic systems of balance laws?

## 4.1 A Class of Position-Dependent Semilinear Systems

Since our error estimates follow from stability theories for the continuous equations, the only relevant setup is the Bressan-Glimm framework of $BV$ solutions. Hence, since we aim at obtaining results relevant for practical applications, we chose to avoid the smallness assumptions on initial data which are necessary for genuinely nonlinear problems, left apart those systems belonging to the Temple class. This choice leaves us with the class of semilinear systems, with a linear convective part, and possible nonlinearities confined to a lower-order right-hand side. This encompasses the systems of two equations (conservation of mass, evolution of the momentum) obtained



D. Amadori and L. Gosse, *Error Estimates for Well-Balanced Schemes on Simple Balance Laws*, SpringerBriefs in Mathematics,
DOI 10.1007/978-3-319-24785-4_4

through the so-called "inertial approximation" of quasilinear hydrodynamics models; such a simplification is often encountered in shallow water studies [13], and for semiconductor simulations [16, 22, 33]. In particular, *treating the temperature as a scalar and ignoring the drift energy are not critical approximations [...], but may become more important with the sharper field profiles in smaller devices* (see [33, p. 738]) and *it is invalid in the vicinity of junctions of one-dimensional devices, or near the contacts of two dimensional ones* (see [22, p. 922]). Restricting our scope to situations where inertial approximations are justified, and assuming that pressure laws are isothermal, which is quite common for semiconductor applications, we are led to consider models written like,

$$\begin{cases} \partial_t \rho + \partial_x J = 0 \\ \partial_t J + \partial_x \rho = 2k(x) g(\rho, J) \end{cases} \tag{4.1}$$

under the assumption that for some $c > 0$,

$$k \in L^1(\mathbb{R}) \cap L^\infty(\mathbb{R}), \qquad k(x) \geq 0 \tag{4.2}$$

$$\partial_J g \leq -c < 0, \qquad |\partial_\rho g| < |\partial_J g|. \tag{4.3}$$

It's convenient to assume that there exists a $C^1$ bounded map $A(\rho)$ such that

$$g(\rho, A(\rho)) = 0 \qquad \text{for all } \rho. \tag{4.4}$$

The curve $J = A(\rho)$ will be called *equilibrium curve*. By taking the derivative of (4.4) and using (4.3), it follows that a so-called *sub-characteristic condition* holds:

$$|A'(\rho)| < 1. \tag{4.5}$$

A typical choice for $g$ is given by the *relaxation term*:

$$g(\rho, J) = A(\rho) - J. \tag{4.6}$$

System (4.1) with assumptions (4.2)–(4.4) perfectly matches the two-scale relaxation framework studied in [12]. A slight variant of (4.6) would be for instance,

$$g(x, \rho, J) = A(x, \rho) - J, \qquad x \mapsto A(x, \cdot) \in C^1(\mathbb{R}), \qquad \sup_x |\partial_\rho A(x, \rho)| < 1.$$

Such a model, which already appears in [14, 29], wouldn't strongly modify both interaction estimates and resulting error bounds, like an elementary semi-conductor model, for which the convective part corresponds to a lattice temperature $\theta_0$,

$$\partial_t J + \partial_x(\theta_0 \rho) = \frac{1}{\tau(x)}\big(\tau(x) E(x) \rho - J\big), \qquad \tau(x)|E(x)| < \sqrt{\theta_0}, \tag{4.7}$$

with $E(x)$ a small static electric field, and $\tau(x)$ standing for a space-dependent relaxation time depending on the local doping concentration. In terms of "microscopic diagonal" variables $f^{\pm}$ (or Riemann invariants), defined by

$$\rho = f^+ + f^-, \qquad J = f^+ - f^-$$

the system (4.1) rewrites as a discrete-velocity kinetic model:

$$\begin{cases} \partial_t(f^-) - \partial_x(f^-) = -k(x)\, G(f^-, f^+) \\ \partial_t(f^+) + \partial_x(f^+) = \phantom{-}k(x)\, G(f^-, f^+) \end{cases} \tag{4.8}$$

where $G(f^-, f^+) := g\left(f^+ + f^-, f^+ - f^-\right)$. Initial data for (4.8) are such that

$$f^{\pm}(t=0,\cdot) = f_0^{\pm} \in L^1 \cap BV(\mathbb{R}). \tag{4.9}$$

Our semi-linear, space-dependent model (4.1) belongs to the class of relaxation systems [23], which was intensively studied both analytically and numerically more a decade ago, mostly for constant coefficients $\partial_x k \equiv 0$, though: see [5, 21, 27, 32, 36], also [6–9, 34] and the survey by Natalini [28].

## 4.2 Main Error Estimate $\mathscr{E}_1$ for Weak Relaxation Regime

### *4.2.1 Statement and First Comments*

**Theorem 4.1** (see [3]) *Assume (4.2), (4.6) and the sub-characteristic condition (4.5); for $x_1 < x_2$ and $2t \le x_2 - x_1$, set*

$$I(t) = \int_{x_1+t}^{x_2-t} |f_{\Delta x}^{\pm}(t,x) - f^{\pm}(t,x)| dx$$

*being $f_{\Delta x}^{\pm}$ defined according to the WB algorithm in Sect. 4.3 (see Fig. 4.3).*

*(i) There exists a positive constant $C > 0$ such that, under a smallness assumption $\|k\|_{L^1(x_1,x_2)} < C$, the following first-order local error estimate holds for $\Delta x$ sufficiently small (see (4.48)),*

$$I(t) \le KI(0) + \Delta x \cdot \mathscr{E}_1, \tag{4.10}$$
$$\mathscr{E}_1(t, x_1, x_2) = \left(2KC_0 + K_0\right) \|k\|_{L^1(x_1,x_2)} + \left(2C_0 - 1\right) \|k\|_{L^1(x_1+t,x_2-t)},$$

*where $C_0$ stands for the Maxwellian gap (see (4.25)) and*

$$K = \frac{C}{C - \|k\|_{L^1(x_1,x_2)}} \geq 1,$$
$$K_0 = 1 + \frac{4K^2}{C(3K+1)} \left(\mathrm{TV}\{f_0^\pm; (x_1, x_2)\} + 2C_0 \|k\|_{L^1(x_1,x_2)}\right).$$

*(ii) If $k \in L^1 \cap BV$, then one has moreover a "half-order", time-dependent estimate,*

$$I(t) \leq I(0) + 2t\sqrt{\Delta x} \cdot \mathscr{E}_2 \tag{4.11}$$

*where,*

$$\mathscr{E}_2(t, x_1, x_2) = \sqrt{C_0 \|k\|_{L^1(x_1,x_2)} A(t)} + \sqrt{\Delta x}\, C_0 \|k\|_{L^1(x_1,x_2)} \|k\|_{L^\infty(x_1,x_2)},$$
$$A(t) = \frac{32}{C_0\, t} \mathrm{TV}\{f_0^\pm; (x_1, x_2)\} + \mathrm{TV}\{k; (x_1, x_2)\}.$$

Some comments on the main estimates (4.10), (4.11) are now in order.

- The value of the constant $C$ in (i) is explicit, being $C = \frac{3}{16} \log(4/3)$: see (4.73).
- The smallness of $\Delta x$ in (i) depends on $\|k\|_\infty$: indeed, (4.48) amounts to require that $\|k\|_\infty \Delta x \leq 8C/3$.
- Thanks to (4.9), the initial error is bounded by $\Delta x \cdot \mathrm{TV}\{f_0^\pm; (x_1, x_2)\}$ as soon as the algorithm is initialized with a convenient sampling of $f_0^\pm$, see (4.70).
- When both the assumptions of the estimates are valid, then one has

$$\boxed{I(t) \leq I(0) + \min\left\{(K-1)I(0) + \Delta x \cdot \mathscr{E}_1;\ 2t\sqrt{\Delta x} \cdot \mathscr{E}_2\right\}.} \tag{4.12}$$

The first estimate in (4.12) is **uniform in time $t$ and first-order in $\Delta x$**, but is meaningful only for "weak relaxation regime", where $K$ remains finite. The complementary estimate is **linear in time $t$ and half-order in $\Delta x$**, which was to be expected as, in strong relaxation regime, the system (4.1) behaves like a scalar conservation law for which optimal convergence order is studied in [31].

## 4.2.2 Strategy of Proof and Algorithmic Implications

The proof of the time-uniform estimate (4.10) relies on an application of the Bressan-Liu-Yang $L^1$-stability theory to an homogeneous but non-conservative, version of system (4.1), see (4.13). A Godunov scheme can be set up, relying on a Riemann solver where the effects of the localized relaxation term are handled by means of a supplementary, static, jump relation, called "standing wave", or "zero-wave". Our estimate (4.12) reveals that *in a context where* $\mathrm{TV}(k)$ *is (locally) big, accurate approximations can be obtained (perhaps beyond a certain time) by means of numerical schemes relying on this type of Riemann solvers, where the source term is handled like a "local scattering center" inducing a stationary discontinuity,* as suggested by Glimm and Sharp in [15]. Besides, the present estimates improve previous ones for semi-linear "locally damped wave equations", see [30, 35],

$$\partial_t \rho + \partial_x J = 0, \qquad \partial_t J + \partial_x \rho = -2k(x)g(J),$$

formerly studied in [4]. For this system, (4.10) holds for any $\Delta x \geq 0$ and provided

$$C_0 = \|g\|_\infty, \qquad C = \frac{1}{4 \cdot Lip(g)}.$$

The only restriction is a small $L^1$ norm for $k$, (equivalently a small variation of $a$)

$$\text{TV } a = \|k\|_{L^1} < \frac{1}{4 \cdot Lip(g)},$$

which is needed in order to ensure that $K < \infty$. Further details are presented in Remark 4.7. This new estimate (4.12) appears like being specific to so-called "well-balanced" methods where source terms are concentrated onto interfacial discontinuities: the fact that $\mathscr{E}_1$ doesn't grow in time and is independent of TV $(k)$, at least when $\|k\|_{L^1}$ is small enough, suggests that this type of algorithms should outperform more conventional (time-splitting, see e.g. [17, 26]) ones when $k(x)$ display strong variations. Accordingly, the time-uniform estimate $\mathscr{E}_1$ is sharper in case the relaxation term is multiplied by a small, but oscillating (or at least, displaying areas of strong variation) coefficient. This meets with early implementations of the so-called "generalized Glimm scheme" by Weinan E [37] in a context of homogenization of scalar balance laws. Such a difference was already seen in [4] on a simpler model of damped wave equation. Moreover, it can give hints on why WB algorithms deliver high accuracy results on shallow water equations in presence of a steep topography: our results don't strictly apply to such a quasi-linear model, though.

## 4.3 Construction of the Well-Balanced Approximation

In this context, the WB approach consists in dealing with the inhomogeneous system (4.8) by means of a non-conservative homogeneous $3 \times 3$ system, which turns out to be equivalent for smooth $a(x)$,

$$\begin{cases} \partial_t \rho + \partial_x J & = 0, \\ \partial_t J + \partial_x \rho - 2g(\rho, J)\, \partial_x a & = 0, \\ \partial_t a & = 0, \end{cases} \qquad a = a(x) \doteq \int_{-\infty}^{x} k(y)\, dy, \tag{4.13}$$

or equivalently, since $G(f^+,f^-) = g\big((f^+ + f^-), (f^+ - f^-)\big)$,

$$\partial_t f^{\mp} \mp \partial_x f^{\mp} \pm G(f^+,f^-)\partial_x a = 0, \qquad \partial_t a = 0. \tag{4.14}$$

From assumption (4.2) one has that

$$a(x) \in BV(\mathbb{R}) \cap C(\mathbb{R}), \qquad a_x \geq 0. \tag{4.15}$$

Characteristic speeds of (4.14) are $\lambda = \{-1, 0, 1\}$ with corresponding eigenvectors

$$\mathbf{r}_- = (0, 1, 0)^t, \qquad \mathbf{r}_0 = (G, G, 1)^t, \qquad \mathbf{r}_+ = (1, 0, 0)^t,$$

where $G(f^+,f^-) := g((f^+ + f^-), (f^+ - f^-))$. The *0-wave curves* are those characteristic curves corresponding to $\lambda = 0$. Characteristic curves for $\lambda = \pm 1$ are straight lines, but for $\lambda = 0$ they are straight lines whenever $A \equiv 0$ (see [4]).

"Handling source terms by means of wave interactions" appears to trace back to Glimm and Sharp [15]. It consists in localizing a source term of bounded extent into a countable collection of "local scattering centers" rendered by Dirac masses, in order to integrate it inside a Riemann solver by means of an elementary (obviously very linearly degenerate) wave. It is extensively used for weakly nonlinear kinetic equations in [18, Part II].

### *4.3.1 First Considerations on the 3 × 3 Riemann Problem*

As usual, let a Riemann data for (4.14)

$$U_\ell = (f_\ell^-, f_\ell^+, a_\ell), \qquad U_r = (f_r^-, f_r^+, a_r)$$

be given. The Riemann problem for system (4.14) is solved in terms of the three characteristic families, resulting in three waves: the two $\pm 1$-waves, with corresponding speed $\pm 1$, where only $f^\pm$ can change its value; and the 0-*wave*, corresponding to the stationary field of (4.14), evolving along the stationary equations

$$\partial_x f^\pm = k(x) G\left(f^-, f^+\right), \tag{4.16}$$

or equivalently, from (4.1):

$$\partial_x J = 0, \qquad \partial_x \rho = 2k(x) g(\rho, J). \tag{4.17}$$

Notice that $J$ is constant along stationary solutions. In terms of the diagonal variables $f^{\pm}$, the equilibrium curve $J = A(\rho)$, i.e. the level curve $G = 0$, is clearly expressed by

$$f^{+} - f^{-} = A\left(f^{+} + f^{-}\right).$$

By (4.5), $|A'| < 1$, and thanks to the implicit function theorem, the corresponding curve realizes a graph in the $f^{\pm}$ plane. Indeed for each $f^{-}$ the map $\mathbb{R} \to \mathbb{R}$,

$$x \mapsto x - f^{-} - A\left(x + f^{-}\right)$$

is strictly increasing and tends to $\pm\infty$ as $x \to \pm\infty$. Hence there exists a function $f^{+} = E(f^{-})$, globally defined on $\mathbb{R}$, along which the source vanishes. This map $E$ is smooth and its derivative equals

$$E'(f^{-}) = \frac{1 + A'(y)}{1 - A'(y)}, \qquad y = E(f^{-}) + f^{-}.$$

Thanks to (4.5), $E' > 0$ thus $E$ is strictly increasing, and the range of $E'$ is $(0, +\infty)$. Now we observe the following interesting feature: from (4.3) it follows that

$$\frac{\partial G}{\partial f^{-}} = \partial_{\rho} g - \partial_{J} g > 0, \qquad \frac{\partial G}{\partial f^{+}} = \partial_{\rho} g + \partial_{J} g < 0 \tag{4.18}$$

and therefore the gradient of $G$ "points" in the bottom-right direction of the $(f^{-}, f^{+})$-plane. Consequently $G > 0$ below the graph and $G < 0$ above the graph (see the arrows on Fig. 4.2). The intermediate states in the Riemann fan are (see Fig. 4.1)

$$U_1 = (f_*^{-}, f_\ell^{+}, a_\ell), \qquad U_2 = (f_r^{-}, f_*^{+}, a_r),$$

while the waves appearing in the solution are as follows: $U_\ell$ and $U_1$ are connected by a $(-1)$-wave of size $\sigma_{-1}$, $U_1$ and $U_2$ are connected by a 0-wave of size $\sigma_0$, and $U_2$ and $U_r$ are connected by a 1-wave of size $\sigma_1$ where

$$\begin{cases} \sigma_{-1} = f_*^{-} - f_\ell^{-} = (f_*^{-} - f_\ell^{+}) - (f_\ell^{-} - f_\ell^{+}) = J_\ell - J_* = \rho_{*,\ell} - \rho_\ell \\ \sigma_0 = a_r - a_l \\ \sigma_1 = f_r^{+} - f_*^{+} = (f_r^{+} - f_r^{-}) - (f_*^{+} - f_r^{-}) = J_r - J_* = \rho_r - \rho_{*,r}. \end{cases} \tag{4.19}$$

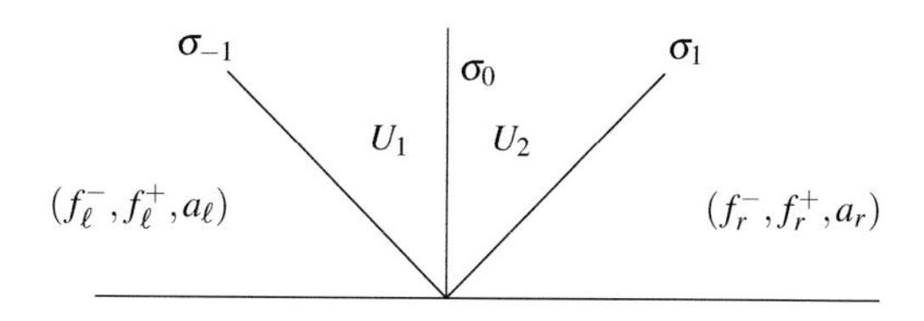


**Fig. 4.1** Schematic view of a Riemann problem for (4.14)

Here the "$*$" signals that the corresponding value is related to the 0-wave: more precisely, $(\rho_{*,\ell}, J_*)$ and $(\rho_{*,r}, J_*)$ denote the left and right states separated by the 0-wave, respectively, in terms of the macroscopic variables $(\rho, J)$.

*Remark 4.1* There exists a practical way to construct the problem for small $\delta$. If $\delta = 0$, there is no zero-wave thus $U_1 = U_2$ is given by the state $P = (f_r^-, f_\ell^+)$, that corresponds to the intersection of the $(-1)$-wave issued from $(f_\ell^-, f_\ell^+)$ and the $(+1)$-wave issued from $(f_r^-, f_r^+)$. Clearly here $J_* = f_\ell^+ - f_r^-$.

For $\delta > 0$ small, the value of $J_*$ can be obtained by perturbation as follows:

- In the very special case where $G(P) = 0$, that is, the intersection point $P$ lies on the equilibrium curve, then the intermediate states $U_1, U_2$ again coincide whatever is the value of $\delta > 0$.
- Assume now that $G(P) > 0$, the other case being similar. Then, for a convenient $J_*$, one has to solve the equation

$$\partial_a \rho = 2g(\rho, J_*), \qquad \rho(a_l) = \rho_{*,\ell}.$$

1. Define $B(\rho, J)$ by integrating up to a constant $\frac{1}{2g}$ with respect to $\rho$,

$$B(\rho, J) = \int^{\rho} \frac{d\rho'}{2g(\rho', J)}. \tag{4.20}$$

For each value of the parameter $J$, the above function is well defined and monotone in a neighborhood of a point $(\bar{\rho}, J)$ such that $g(\bar{\rho}, J) \neq 0$. Then, the left and right states of the 0-wave satisfy the relation

$$B(\rho_{*,r}, J_*) - B(\rho_{*,\ell}, J_*) = \int_{\rho_{*,\ell}}^{\rho_{*,r}} \frac{d\rho'}{2g(\rho', J)} = a_r - a_\ell.$$

Notice that, even if $B$ is defined by (4.20) up to a function depending on $J$, the above difference does not depend on the choice of the particular function.

2. Now, by the very definition of $f^\pm$, we have

$$\rho_{*,\ell} + J_* = 2f_\ell^+, \qquad \rho_{*,r} - J_* = 2f_r^- \tag{4.21}$$

then, by taking advantage of the fact that $f^+$ (resp. $f^-$) is constant across a $-1$-wave (resp. across a 1-wave), we can write an implicit equation for $J_*$, in terms of the parameters $f_\ell^+$ and $f_r^-$:

$$B\left(2f_r^- + J_*, J_*\right) - B\left(2f_\ell^+ - J_*, J_*\right) = a_r - a_\ell. \tag{4.22}$$

The equation (4.22) already appeared in the context of diffusive numerical approximations, in a slightly different form: see the book [18], page 150.

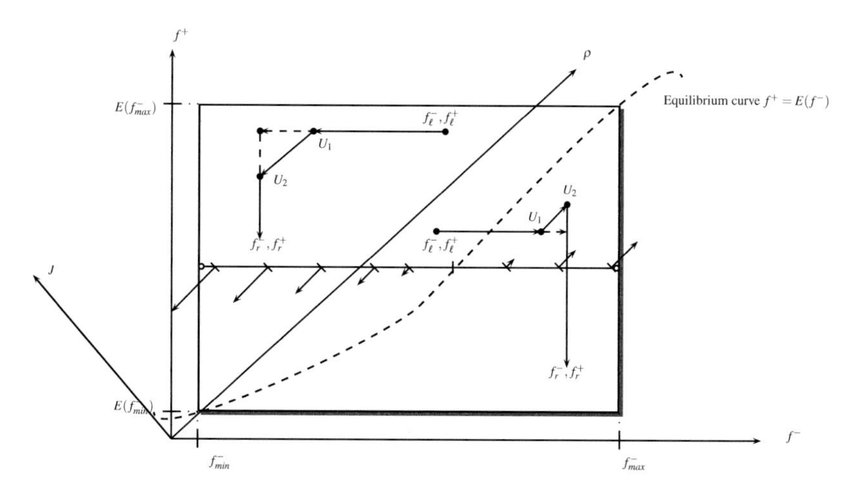


**Fig. 4.2** Invariant domain for 3 × 3 system (4.14) and 2 sets of initial/final states. *Diagonal arrows* stand for the projection of the vector $\mathbf{r}_0$ on the $f^{\pm}$ plane

### *4.3.2 Shape of Positively Invariant Domains*

A standard roadmap for establishing control on the amplitude of approximations is to seek a positively invariant domain for the Riemann problem: see Fig. 4.2.

**Proposition 4.1** *Let* $f^-_{min} < f^-_{max}$, $P_1 = \big(f^-_{min}, E(f^-_{min})\big)$, $P_2 = \big(f^-_{max}, E(f^-_{max})\big)$, *and* $\delta := a_r - a_\ell > 0$. *Then the rectangle*

$$D = [f^-_{min}, f^-_{max}] \times [E(f^-_{min}), E(f^-_{max})] \tag{4.23}$$

*is positively invariant for the unique solution of the Riemann problem. More precisely, for any pair of states* $(f^-_\ell, f^+_\ell)$ *and* $(f^-_r, f^+_r) \in D$ *and for* $\delta = a_r - a_\ell > 0$, *there exists a single choice of the intermediate states* $U_1$, $U_2$ *for which one has*

$$\big| |f^-_r - f^-_\ell| - |\sigma_{-1}| \big| \le C_0\delta, \qquad \big| |f^+_r - f^+_\ell| - |\sigma_1| \big| \le C_0\delta \tag{4.24}$$

*where* $C_0$ *measures the "Maxwellian gap" in* $L^\infty$:

$$C_0 = \max\{|G(f^-, f^+)|;\ (f^-, f^+) \in D\}. \tag{4.25}$$

*Proof* 1. Thanks to (4.21), all the intermediate states in the Riemann problem can be deduced from $J_*$, and values $f^+_*$ and $f^-_*$ are defined by the identity

$$f^+_* - f^-_r = f^+_\ell - f^-_* = J_*.$$

Clearly, if $G(f_r^-, f_\ell^+) = 0$, then $J_* = f_\ell^+ - f_r^-$ for every $\delta > 0$. But when $G(f_r^-, f_\ell^+) \neq 0$, the value of $\tilde{J}$ is implicitly defined by (4.22): indeed let $(f^-, f^+) \in D$ be such that $G(f^-, f^+) > 0$ (that is, $(f^-, f^+)$ **below** the equilibrium curve, the opposite case being similar) and define a function $F$ as follows:

$$(J, \delta) \mapsto F(J, \delta; f^\pm) = B\left(2f^- + J, J\right) - B\left(2f^+ - J, J\right) - \delta$$
$$= \int_{2f^+ - J}^{2f^- + J} \frac{d\rho'}{2g(\rho', J)} - \delta \tag{4.26}$$

(subscripts in $f^\pm$ were dropped). One easily finds a particular solution for $\delta = 0$,

$$F(J_0, 0; f^\pm) = 0 \Leftrightarrow J_0 = f^+ - f^-.$$

This solution corresponds to the case where there is no zero-wave because $\delta = a_r - a_\ell$ vanishes. Let us verify that the following property holds:

$$0 \neq \frac{\partial F}{\partial J}(J_0, 0; f^\pm) = \left(\partial_J B + \partial_\rho B\right)(2f^- + J_0, J_0) - \left(\partial_J B - \partial_\rho B\right)(2f^+ - J_0, J_0),$$

but since $2f^+ - J_0 = f^+ + f^- = 2f^- + J_0$, this expression reduces to

$$\frac{\partial F}{\partial J}(J_0, 0; f^\pm) = 2\partial_\rho B(f^+ + f^-, f^+ - f^-) = \frac{1}{G(f^+, f^-)} \neq 0.$$

The implicit functions theorem ensures existence and uniqueness of a smooth,

$$\tilde{J} : (0, \varepsilon) \times D \to \mathbb{R}, \qquad \delta, f^\pm \mapsto \tilde{J}(\delta; f^\pm), \tag{4.27}$$

such that, for $0 < \varepsilon \ll 1$ and $G(f^+, f^-) \neq 0$, $J(0; f^\pm) = J_0 = f^+ - f^-$ and

$$F(\tilde{J}, \delta; f^\pm) = 0 \quad \Leftrightarrow \quad \int_{2f^+ - \tilde{J}}^{2f^- + \tilde{J}} \frac{d\rho'}{2g(\rho', \tilde{J})} = \delta. \tag{4.28}$$

Moreover, since $\partial F / \partial \delta = -1$, we make explicit the derivative of $\tilde{J}$:

$$\frac{\partial \tilde{J}}{\partial \delta} = \frac{1}{\frac{\partial F}{\partial J}(\tilde{J}, \delta; f^\pm)}, \qquad \frac{\partial \tilde{J}}{\partial \delta}(\delta = 0; f^\pm) = G(f^+, f^-) > 0.$$

Under those smallness restrictions and $(f^-, f^+)$ being fixed below the equilibrium curve, the restriction $\delta \mapsto \tilde{J}$ is increasing because $G(f^+, f^-) > 0$. Moreover, as soon as $\tilde{J}(\delta, f^\pm)$ is defined, the segment in the state space $\rho, J$ along which the integral in (4.26) is computed, parametrized by $\rho$ as follows,

$$[2f^+ - \tilde{J}, 2f^- + \tilde{J}] \ni \rho \mapsto (\rho, \tilde{J}) \tag{4.29}$$

does not intersect the equilibrium curve: otherwise, the integral in (4.28) would blow up, instead of being equal to $\delta$.

2. Now we intend to verify that $\frac{\partial F}{\partial J}(\tilde{J}, \delta; f^\pm) > 0$, in order to establish that $\tilde{J}$ is indeed increasing as soon as it is defined. Using the definition of $B$, see (4.20),

$$\begin{aligned}\frac{\partial F}{\partial J}(J, \delta; f^\pm) &= \big(\partial_J B + \partial_\rho B\big)(2f^- + J, J) - \big(\partial_J B - \partial_\rho B\big)(2f^+ - J, J)\\ &= \int_{2f^+-J}^{2f^-+J} \frac{1}{2g^2(\rho', J)} |\partial_J g(\rho', J)|\, d\rho' + \frac{1}{2g(2f^- + J, J)}\\ &\quad + \frac{1}{2g(2f^+ - J, J)}.\end{aligned} \tag{4.30}$$

Assuming again that $G(f^+, f^-) > 0$, if $J$ is set to $J = \tilde{J}(\delta; f^\pm)$ then the extrema of the integral above satisfy $2f^+ - \tilde{J} < 2f^- + \tilde{J}$ (see Fig. 4.2). Hence we can take advantage of the last condition in (4.3),

$$\begin{aligned}\int_{2f^+-J}^{2f^-+J} \frac{1}{g^2(\rho', J)} |\partial_J g(\rho', J)|\, d\rho' &\geq \int_{2f^+-J}^{2f^-+J} \frac{1}{g^2(\rho', J)} |\partial_\rho g(\rho', J)|\, d\rho'\\ &\geq \int_{2f^+-J}^{2f^-+J} \frac{1}{g^2(\rho', J)} \partial_\rho g(\rho', J)\, d\rho'\\ &= -\frac{1}{g(2f^- + J, J)} + \frac{1}{g(2f^+ - J, J)}.\end{aligned}$$

We can therefore estimate from below the integral in (4.30) and get

$$\frac{\partial F}{\partial J}(J, \delta; f^\pm) \geq \frac{1}{g(2f^+ - J, J)} > 0.$$

We now deduce that the function $\delta \mapsto \tilde{J}(\delta; f^\pm)$ is actually defined on $\mathbb{R}^+$:

- There exists $J_{\max}$ such that the interval in (4.29) has no intersection with the equilibrium curve for $J_0 \leq J < J_{\max}$.
- For $J = J_{\max}$ there exists a point of the interval, $(\bar{\rho}, J_{\max})$, such that $g(\bar{\rho}, J_{\max}) = 0$. Since $g$ is $C^1$, then $g(\rho, J_{\max}) = O(1)(\rho - \bar{\rho})$ and therefore the corresponding integral is not finite.

Hence $\tilde{J}(\delta)$ is defined for every $\delta > 0$, and $\tilde{J}(\delta) \to J_{\max}$ as $\delta \to \infty$. Analogously $\delta \mapsto \tilde{J}(\delta; f^\pm)$ is decreasing when $(f^-, f^+)$ is **above** the equilibrium curve, and that it is defined for all $\delta > 0$ finite. The monotonicity of $\tilde{J}$ implies that the domain $D$ is positively invariant for the Riemann problem, see Fig. 4.2.

3. Finally, concerning (4.24), we use (4.16) and (4.19) to estimate the jump in the $f^\pm$ coordinate across the 0-wave, that is:

$$|f_*^+ - f_\ell^+| = |f_r^- - f_*^-| \le \sup_D |G| \cdot \delta$$

and this yields,

$$\big| |f_r^+ - f_\ell^+| - |\sigma_1| \big| = \big| |f_r^+ - f_\ell^+| - |f_r^+ - f_*^+| \big| \le |f_*^+ - f_\ell^+| \le C_0 \delta$$

with $C_0$ as (4.25). An analogous estimate holds for $\sigma_{-1}$ so (4.24) holds.

### 4.3.3 Total Variation Estimate of the WB Approximation

Let $D$ be a rectangle as in (4.23) that contains the values of initial data $(f_0^-, f_0^+)$. By means of Proposition 4.1, since $a_x \ge 0$, up to a suitable choice of the initial data, the approximate solution remains confined inside the region $D$

$$\forall t > 0, \qquad (f^-, f^+)(t, .) \in D. \tag{4.31}$$

Previous results on positively invariant domains for the $3 \times 3$ Riemann problem for (4.14) allow to easily derive uniform bounds on the total variation of the corresponding WB approximation thanks to its peculiar structure. The method hereafter is taken from [20, p. 643], and we recall it now for completeness:

- Differentiate in time each equation on $f^\mp$ in (4.14), multiply by $sgn(\partial_t f^\mp)$ and then integrate on $x \in \mathbb{R}$. It comes

$$\forall t > 0, \qquad \partial_t \int_{\mathbb{R}} \big( |\partial_t f^+(t,x)| + |\partial_t f^-(t,x)| \big) \cdot dx \le 0,$$

  thanks to the quasi-monotonicity of $G$.
- An easy observation is that, by taking moduli, it comes:

$$|\partial_x f^\pm| - |G(f^-, f^+)\partial_x a| \le |\partial_t f^\pm| \le |\partial_x f^\pm| + |G(f^-, f^+)\partial_x a|.$$

- It remains to integrate in space in order to obtain the estimate:

$$\int_{\mathbb{R}} \big( |\partial_x f^+(t,x)| + |\partial_x f^-(t,x)| \big) \cdot dx$$
$$\le \int_{\mathbb{R}} \big( |\partial_x f^+(0,x)| + |\partial_x f^-(0,x)| \big) \cdot dx + 4 C_0 \mathrm{TV}\ a$$

  where $C_0$ is given as in (4.25).

Recall also that TV $a = \|k\|_{L^1}$. The last point to clarify addresses the fact that we seek a BV-bound on an *approximation* depicted on Fig. 4.3, and not the exact solutions $f^\pm$ of (4.14). However, since we chose to work with the Courant number

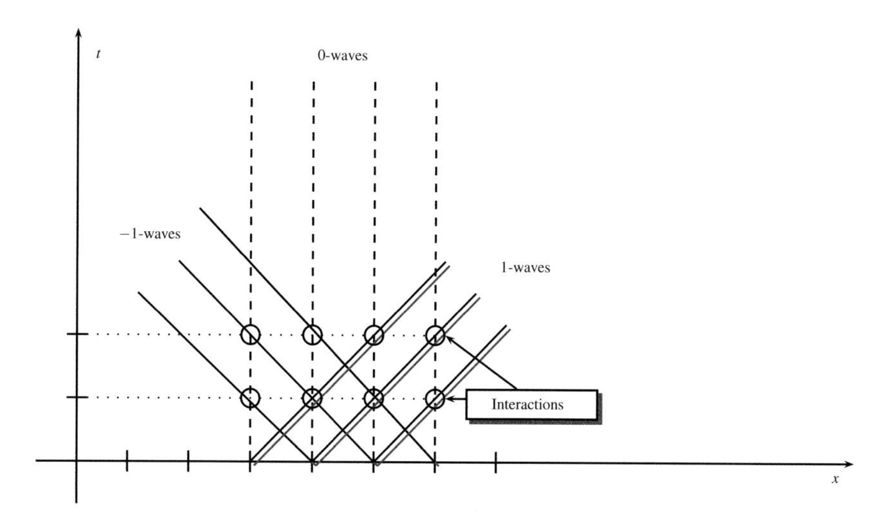


**Fig. 4.3** Schematic view of a WB approximate solution: circles indicate Riemann problems studied in Proposition 4.1. Since the Courant number is 1, constant states always lie in between them

equal to 1, the only difference between the WB approximation and the exact solutions lies in the sampling of initial data. Hence one gets the following estimate that does **not depend on time** for the WB approximation:

$$\boxed{\mathrm{TV}\, f^{+}(t,\cdot) + \mathrm{TV}\, f^{-}(t,\cdot) \leq \mathrm{TV}\, f^{+}(0,\cdot) + \mathrm{TV}\, f^{-}(0,\cdot) + 4\, C_0\, \|k\|_{L^1}.} \tag{4.32}$$

This BV-bound matches the one obtained in [4] with rather different methods.

## 4.4 Refined Scattering Interaction Estimates

In this section we assume for simplicity that (4.6) and (4.5) hold, that is $g(\rho, J) = A(\rho) - J$ with $|A'| < 1$. The extension to more general relaxation terms (4.2)–(4.4) follows without substantial difficulties, but at the price of tedious computations. We first study interactions between various patterns of waves for the system (4.14) in order to, in a second step, estimate the time-variation of the Lyapunov $L^1$-functional.

### 4.4.1 A Lemma Based on Sub-characteristic Condition

We start with a Gronwall-type lemma that exploits the sub-characteristic condition (4.5). Since $|A'| < 1$ and $\rho$ ranges over a compact set, there exists a positive constant $\alpha < 1$ such that $\boxed{|A'| \leq \alpha}$.

**Lemma 4.1** *Let $|A'| \leq \alpha$ and let $\rho(a, J)$ satisfy the parameter-dependent differential equation*

$$\forall J \in Range(\tilde{J}), \qquad \frac{\partial \rho(a, J)}{\partial a} = 2\Big(A(\rho) - J\Big), \qquad a \in [a_\ell, a_r].$$

*For some $J_2 > J_1$, $a \in [a_\ell, a_r]$, define $\phi(a) = \rho(a, J_2) - \rho(a, J_1)$, and assume that*

$$\phi(a_r) = \rho(a_r, J_2) - \rho(a_r, J_1) = J_2 - J_1 > 0. \tag{4.33}$$

*Then the following inequalities hold:*

$$\forall a \in [a_\ell, a_r], \qquad \phi(a_r) - \phi(a) \leq -\widetilde{c}\,(J_2 - J_1)\,(a_r - a) \tag{4.34}$$

*with*

$$\widetilde{c} = (1 - \alpha)\frac{e^{2\alpha(a_\ell - a_r)} - 1}{\alpha(a_\ell - a_r)} > 0, \tag{4.35}$$

*and*

$$\forall a \in [a_\ell, a_r], \qquad 0 < \phi(a) - \phi(a_r) \leq \widetilde{C}\,(J_2 - J_1)\,(a_r - a) \tag{4.36}$$

*with*

$$\widetilde{C} = (1 + \alpha)\frac{e^{2\alpha(a_r - a_\ell)} - 1}{\alpha(a_r - a_\ell)}. \tag{4.37}$$

*Proof* As soon as $\phi(a) > 0$ (which is true by continuity for $a$ close to $a_r$), $\phi$ satisfies

$$\begin{aligned}\phi'(a) &= 2\,(A(\rho(a, J_2)) - A(\rho(a, J_1)) - 2\,(J_2 - J_1)\\ &\leq 2\alpha\phi(a) - 2(J_2 - J_1).\end{aligned} \tag{4.38}$$

An application of the Gronwall lemma yields

$$e^{2\alpha(a - a_r)}\phi(a_r) - \phi(a) \leq \frac{J_2 - J_1}{\alpha}\Big(e^{2\alpha(a - a_r)} - 1\Big). \tag{4.39}$$

By summing, subtracting and then using (4.39) and (4.33), we infer that

$$\begin{aligned}
\phi(a_r) - \phi(a) &= \left[e^{2\alpha(a-a_r)}\phi(a_r) - \phi(a)\right] - \left(e^{2\alpha(a-a_r)} - 1\right)\phi(a_r) \\
&\le \frac{J_2 - J_1}{\alpha}\left(e^{2\alpha(a-a_r)} - 1\right) - \left(e^{2\alpha(a-a_r)} - 1\right)\phi(a_r) \\
&= (J_2 - J_1)\,\frac{e^{2\alpha(a-a_r)} - 1}{\alpha}\,(1-\alpha) \\
&= -(a_r - a)(J_2 - J_1)\left\{\frac{e^{2\alpha(a-a_r)} - 1}{\alpha(a-a_r)}\right\}(1-\alpha) \\
&\le -\widetilde{c}(a_r - a)(J_2 - J_1)
\end{aligned}$$

with $\widetilde{c}$ as in (4.35). This proves (4.34). Such inequality, rewritten as

$$\phi(a_r) + (a_r - a)\,(J_2 - J_1)\,\widetilde{c}\,(1-\alpha) \le \phi(a), \qquad a \in [a_\ell, a_r],$$

shows that $\phi$ remains positive, so the above argument holds as soon $\phi(a)$ is defined. To prove (4.36), we compute again $\phi'$ and find the opposite inequality to (4.38):

$$\phi'(a) \ge -2\alpha\phi(a) - 2(J_2 - J_1),$$

where we used also that $\phi(a) > 0$. The Gronwall lemma yields

$$e^{-2\alpha(a-a_r)}\phi(a_r) - \phi(a) \ge -\frac{J_2 - J_1}{\alpha}\left(e^{-2\alpha(a-a_r)} - 1\right).$$

By proceeding as in the first part of the proof, we obtain

$$\begin{aligned}
\phi(a_r) - \phi(a) &= \left[e^{2\alpha(a_r-a)}\phi(a_r) - \phi(a)\right] - \left(e^{2\alpha(a_r-a)} - 1\right)\phi(a_r) \\
&\ge -\frac{J_2 - J_1}{\alpha}\left(e^{2\alpha(a_r-a)} - 1\right) - \left(e^{2\alpha(a_r-a)} - 1\right)\phi(a_r) \\
&= -(J_2 - J_1)\left\{\frac{e^{2\alpha(a_r-a)} - 1}{\alpha(a_r-a)}\right\}(1+\alpha)\,(a_r - a) \\
&\ge -\widetilde{C}(J_2 - J_1)(a_r - a)
\end{aligned}$$

with $\widetilde{C}$ as in (4.37). It remains to change sign in the inequality above to get (4.36).

*Remark 4.2* The function $\phi(a)$ quantifies the dependence of $\rho$ with respect to the parameter $J$. Accordingly, one can notice that, formally,

$$\phi(a) \simeq \frac{\partial\rho}{\partial J}(J_2 - J_1), \qquad \phi(a_r) - \phi(a) \simeq \frac{\partial^2\rho}{\partial J\,\partial a}(J_2 - J_1)(a_r - a).$$

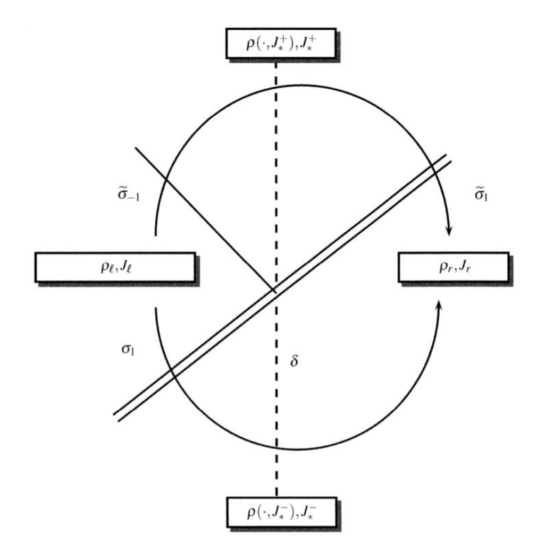


**Fig. 4.4** Interaction pattern corresponding to the scattering of a linear wave $\sigma_1$ by a source-term discontinuity of size $\delta = a_r - a_\ell > 0$

Hence, assuming all the necessary smoothness, estimates (4.34) and (4.36) state in a rigorous manner the informal statement that the mixed derivative is strictly negative,

$$-\widetilde{C} \leq \frac{\partial^2 \rho}{\partial a\, \partial J} \leq -\widetilde{c} < 0.$$

**Lemma 4.2** *Consider the elementary interaction pattern displayed on Fig. 4.4, with* $\delta = a_r - a_\ell > 0$*: the conservation law holds,*

$$\boxed{|\widetilde{\sigma}_1| + |\widetilde{\sigma}_{-1}| = |\sigma_1|.} \tag{4.40}$$

*The reflected wave has always opposite sign with respect to the transmitted one:*

$$\widetilde{\sigma}_1 \cdot \widetilde{\sigma}_{-1} \leq 0. \tag{4.41}$$

*Moreover, the amplitude of the reflected wave is bounded by*

$$|\widetilde{\sigma}_{-1}| \leq \widetilde{C}_1\, \delta |\sigma_1|, \qquad \widetilde{C}_1 = \frac{\widetilde{C}}{2} = (1+\alpha)\frac{e^{2\alpha\delta} - 1}{2\alpha\delta}, \tag{4.42}$$

$\widetilde{C}$ *as in (4.37). The case of a* $(-1)$*-wave interacting with the* 0*-wave is analogous.*

Such a lemma expresses a strong conservation law for the scattering process where an incoming wave $\sigma_1$ is scattered by a zero-wave $\delta$ giving birth to reflected/transmitted waves $\widetilde{\sigma}_{\pm 1}$.

*Proof* It splits into several steps.

- Let $a \mapsto \rho(a, J)$, $a \in [a_\ell, a_r]$, stand for solutions of the ODE problem along the 0-wave associated with a flux value $J$. More precisely, we have, before and after interaction, respectively:

$$\frac{\partial}{\partial a}\rho(a, J_*^-) = 2\left(A(\rho^-) - J_*^-\right), \qquad \rho^-(a_r, J_*^-) = \rho_r,$$
$$\frac{\partial}{\partial a}\rho(a, J_*^+) = 2\left(A(\rho^+) - J_*^+\right), \qquad \rho(a_r, J_*^+) = \rho_r - \widetilde{\sigma}_1.$$

  Notice also that $J_*^- = J_r$, and so $J_*^- - J_*^+ = \widetilde{\sigma}_1$.
- By equating $J_r - J_\ell$ before and after the interaction, we find that

$$\widetilde{\sigma}_1 - \widetilde{\sigma}_{-1} = \sigma_1. \tag{4.43}$$

  On the other hand, by equating $\rho_r - \rho_\ell$ before and after the interaction, and by using the definition of the size of waves, (4.19) in terms of jumps of $\rho$, we get

$$\begin{aligned}\rho_r - \rho_\ell &= \sigma_1 + \left(\rho(a_r, J_*^-) - \rho(a_\ell, J_*^-)\right) && \text{(lower curved arrow on Fig. 4.4)},\\ &= \widetilde{\sigma}_1 + \widetilde{\sigma}_{-1} + \left(\rho(a_r, J_*^+) - \rho(a_\ell, J_*^+)\right) && \text{(upper curved arrow on Fig. 4.4)}.\end{aligned}$$

  Henceforth, one deduces:

$$\widetilde{\sigma}_1 + \widetilde{\sigma}_{-1} + \left(\rho(a_r, J_*^+) - \rho(a_\ell, J_*^+)\right) = \sigma_1 + \left(\rho(a_r, J_*^-) - \rho(a_\ell, J_*^-)\right). \tag{4.44}$$

  By setting

$$\phi(a) = \rho(a, J_*^-) - \rho(a, J_*^+)$$

  and by subtracting (4.43) from (4.44), it comes

$$\begin{aligned}2\widetilde{\sigma}_{-1} &= \left(\rho(a_r, J_*^-) - \rho(a_\ell, J_*^-)\right) - \left(\rho(a_r, J_*^+) - \rho(a_\ell, J_*^+)\right)\\ &= \phi(a_r) - \phi(a_\ell).\end{aligned} \tag{4.45}$$

- Now we claim that

$$-\widetilde{C}\,\delta\,(\widetilde{\sigma}_1)^2 \le 2\widetilde{\sigma}_{-1}\widetilde{\sigma}_1 \le -\widetilde{c}\,\delta\,(\widetilde{\sigma}_1)^2. \tag{4.46}$$

  from which one easily deduces that (4.41) holds: $\widetilde{\sigma}_{-1}\widetilde{\sigma}_1 \le 0$, along with,

$$2|\widetilde{\sigma}_{-1}||\widetilde{\sigma}_1| \le \widetilde{C}\,\delta\,(\widetilde{\sigma}_1)^2,$$

  that gives (4.42). To prove (4.46), we make use of Lemma 4.1.

- Assume first that $\widetilde{\sigma}_1 > 0$: since $\phi(a_r) = J_*^- - J_*^+ = \widetilde{\sigma}_1 > 0$, then the estimates (4.34), (4.36) for $a = a_\ell$ lead to

$$-\widetilde{C}\,\delta\,\widetilde{\sigma}_1 \le \phi(a_r) - \phi(a_\ell) \le -\widetilde{c}\,\delta\,\widetilde{\sigma}_1.$$

  As a result, (4.46) holds for $\widetilde{\sigma}_1 > 0$.
- Oppositely, if $\widetilde{\sigma}_1 < 0$, Lemma 4.1 applies with $\tilde{\phi}(a) = \rho(a, J_*^+) - \rho(a, J_*^-)$:

$$-\widetilde{C}\,\delta\,|\widetilde{\sigma}_1| \,\le\, \tilde{\phi}(a_r) - \tilde{\phi}(a_\ell) \,\le\, -\delta\,|\widetilde{\sigma}_1|\,\widetilde{c},$$

  leading to

$$-\,\widetilde{C}\,\delta\,\widetilde{\sigma}_1 \ge \phi(a_r) - \phi(a_\ell) \ge -\delta\,\widetilde{\sigma}_1\,\widetilde{c}. \tag{4.47}$$

  Multiplying by $\widetilde{\sigma}_1$ in (4.47), we get (4.46) for $\widetilde{\sigma}_1 < 0$.

Therefore the proof of (4.46) is complete, and hence the proof of the Lemma.

### *4.4.2 Accurate Interaction Estimates for WB Approximations*

Lemma 4.2 allows to consider more intricate interaction patterns:

**Proposition 4.2** *Let $U_\ell$ and $U_m$ be connected by a complete Riemann pattern of size $q_{\pm 1}^-$ and $q_0$. Let $U_m$ and $U_r$ be connected by a single wave as described in the cases below. Finally let $q_{\pm 1}^+$ be the sizes of the $\pm 1$-waves solving the Riemann problem for $U_\ell$, $U_r$ (see Figs. 4.5 and 4.6). Under the hypotheses of Proposition 4.1 and for*

$$2\|k\|_\infty\,\Delta x \,\le\, \log\left(\frac{3}{2}\right), \qquad C_1 = \frac{4}{3\log(3/2)} \simeq 3.29 \tag{4.48}$$

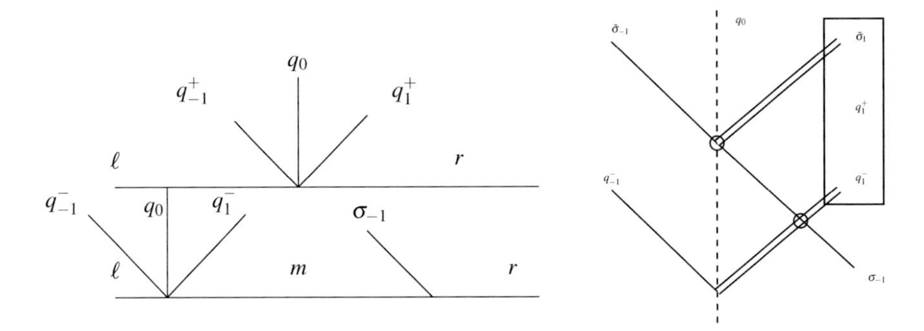


**Fig. 4.5** Illustration of Case (a)

*then the following properties hold:*

*(a) If $U_m$ and $U_r$ be connected by a $-1$-wave of size $\sigma_{-1}$, then*

$$|q_{-1}^{+} - q_{-1}^{-} - \sigma_{-1}| = |q_1^{+} - q_1^{-}| \leq C_1\, q_0\, |\sigma_{-1}|. \tag{4.49}$$

*(b) If $U_m$ and $U_r$ be connected by a 0-wave of size $\sigma_0$, then*

$$|q_{-1}^{+} - q_{-1}^{-}| = |q_1^{+} - q_1^{-}| \leq C_1\, |q_1^{-}|\, \sigma_0. \tag{4.50}$$

*(c) If $U_m$ and $U_r$ be connected by a 1-wave of size $\sigma_1$, then*

$$q_{-1}^{+} = q_{-1}^{-}, \qquad q_1^{+} = q_1^{-} + \sigma_1.$$

The explicit constant $C_1$, in (4.48), was the main reason for setting up Lemma 4.1.

*Proof* Denote $J_*^{-}$, $J_*^{+}$ the intermediate values of $J$ in the Riemann problem for $(U_\ell, U_m)$ and $(U_\ell, U_r)$ respectively. Then the following identities hold:

$$\begin{cases} q_{-1}^{+} - q_{-1}^{-} &= J_*^{-} - J_*^{+}, \\ q_1^{+} - q_1^{-} &= \left(J_*^{-} - J_*^{+}\right) + (J_r - J_m)\,. \end{cases} \tag{4.51}$$

Indeed, recall definitions (4.19): for instance, $q_{-1}^{+} - q_{-1}^{-} = (J_\ell - J_*^{+}) - (J_\ell - J_*^{-})$ yields the first identity (similar for the second). We proceed by increasing difficulty.

1. **Case (c)**. One has $J_r - J_m = \sigma_1$ and $J_*^{-} = J_*^{+}$. Hence the claim simply follows from (4.51), being $q_1^{+} - q_1^{-} - \sigma_1 = 0 = q_{-1}^{+} - q_{-1}^{-}$.
2. **Case (a)**. Recalling sizes (4.19), $\sigma_{-1} = J_m - J_r$, so identities (4.51) lead to

$$q_1^{+} - q_1^{-} = q_{-1}^{+} - q_{-1}^{-} - \sigma_{-1}. \tag{4.52}$$

- Let us proceed by letting both the linear waves $\sigma_{-1}$ and $q_1^{-}$ cross each other (without changing their size). Later, $\sigma_{-1}$ interacts with $q_0$: denote by $\widetilde{\sigma}_{\pm 1}$ the resulting waves so that the final sizes $q_{\pm 1}^{+}$ satisfy

$$q_{\pm 1}^{+} = \widetilde{\sigma}_{\pm 1} + q_{\pm 1}^{-} \qquad \Rightarrow \qquad q_1^{+} - q_1^{-} = \widetilde{\sigma}_1 = q_{-1}^{+} - q_{-1}^{-} - \sigma_{-1} = \widetilde{\sigma}_{-1} - \sigma_{-1}.$$

  Accordingly, equality (4.52) rewrites $q_1^{+} - q_1^{-} = \widetilde{\sigma}_1 = \widetilde{\sigma}_{-1} - \sigma_{-1}$. Applying (4.42) in Lemma 4.2 we get

$$|\widetilde{\sigma}_1| \leq \widetilde{C}_1\, q_0 |\sigma_{-1}|,$$

  so (4.49) holds with

$$C_1 \geq \widetilde{C}_1(q_0) \qquad \forall\, q_0.$$

The choice of the constant $C_1$ will be finalized in the next Case (b).

3. **Case (b)**. Here, the scattering processes related to 2 distinct zero-waves, of sizes $q_0$ and $\sigma_0$ respectively, are "glued" altogether into a unique one. In a linear context, this can be processed by means of "Redheffer products" [19].

   - In this last case we have $J_r = J_m$ and hence (4.51) reduces to

   $$q_1^+ - q_1^- = q_{-1}^+ - q_{-1}^- = J_*^- - J_*^+, \tag{4.53}$$

   which already yields the left part of (4.50).
   - Without loss of generality, one can safely assume that $q_{-1}^- = 0$: indeed, let us show that the seemingly more intricate case $q_{-1}^- \neq 0$ simply reduces to it. Let $\tilde{q}_{\pm 1}^+$ be the result of the reduced interaction involving only $q_0, q_1^-, \sigma_0$; estimates involve only quantities $\tilde{q}_1^+ - q_1^-$ and $\tilde{q}_{-1}^+$. Now, if $q_{-1}^- \neq 0$, then by linearity, resulting waves $q_{\pm 1}^+$ as in Fig. 4.6 satisfy

   $$q_1^+ = \tilde{q}_1^+, \qquad q_{-1}^+ = \tilde{q}_{-1}^+ + q_{-1}^-.$$

   - Accordingly, in situations like in Fig. 4.6 with $\sigma_1 = q_1^-$, Lemma 4.2 gives,

   $$\text{sgn}(\sigma_1^+) = \text{sgn}(\sigma_1) = -\text{sgn}(\sigma_2).$$

   By induction, this property propagates at all scattering events: $\forall n \in \mathbb{N}$,

   $$\text{sgn}(\sigma_n^+) = \text{sgn}(\sigma_n) = -\text{sgn}(\sigma_{n+1}), \ \ \text{sgn}(\sigma_{2n+1}^+) = \text{sgn}(\sigma_1) = -\text{sgn}(\sigma_{2n}^+).$$

   Next, consider quadratic interactions on the right zero-wave (of size $\sigma_0$):

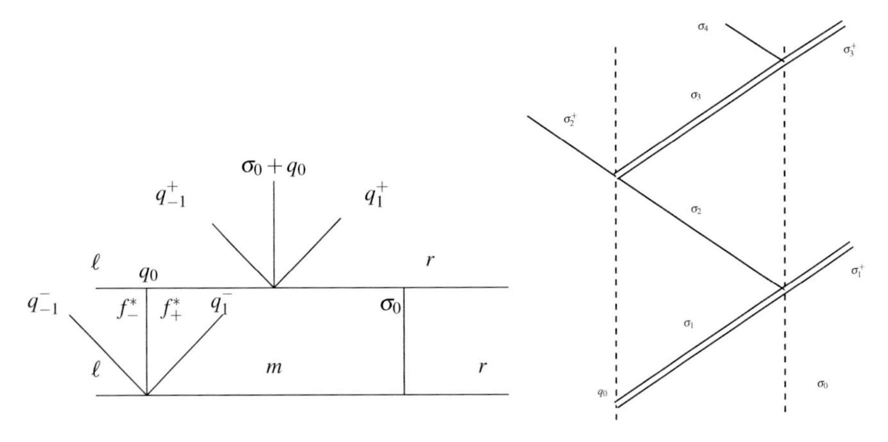


**Fig. 4.6** Illustration of Case (b)

$$|\sigma_{2n+2}| \leq \widetilde{C}_1(\sigma_0)\sigma_0\,|\sigma_{2n+1}| \leq \widetilde{C}_1(\sigma_0)\widetilde{C}_1(q_0)\sigma_0 q_0\,|\sigma_{2n}|$$

where $\widetilde{C}_1$ is as in (4.42), and one has

$$\widetilde{C}_1(x) = (1+\alpha)\frac{\exp(2\alpha x)-1}{2\alpha x} \simeq (1+\alpha) \qquad \text{as } x \to 0.$$

Also, we can estimate both $\sigma_0$, $q_0$ as follows: $\sigma_0, q_0 \leq \Delta x\|k\|_\infty$. Hence

$$|\sigma_{2n+2}| \leq \gamma\,|\sigma_{2n}|, \qquad \gamma \doteq (\widetilde{C}_1(\bar{x})\bar{x})^2, \qquad \bar{x} \doteq \Delta x \cdot \|k\|_\infty$$

and $\gamma \to 0$ as $\Delta x \to 0$ in weak relaxation regime. This immediately implies,

$$|\sigma_{2n+2}| \;\leq\; \gamma^n|\sigma_2| \;\leq\; \gamma^n\widetilde{C}_1(\bar{x})\,|q_1^-|\cdot|\sigma_0|.$$

- It now remains to sum all the even terms:

$$\begin{aligned}\left|\sum_{n=1}^{\infty}\sigma_{2n}^+\right| &= \sum_{n=1}^{\infty}|\sigma_{2n}^+| \leq \sum_{n=1}^{\infty}|\sigma_{2n}| \\ &\leq \left(\sum_{n=1}^{\infty}\gamma^{n-1}\right)\widetilde{C}_1(\bar{x})\,|q_1^-|\cdot|\sigma_0| \;=\; \underbrace{\left(\frac{\widetilde{C}_1(\bar{x})}{1-\gamma}\right)}_{=C_1}|q_1^-|\cdot|\sigma_0|.\end{aligned}$$

To estimate the above defined constant $C_1$, we assume that $\bar{x}$ satisfies

$$\widetilde{C}_1(\bar{x})\bar{x} = (1+\alpha)\frac{\exp(2\alpha\bar{x})-1}{2\alpha} \;\leq\; \frac{1}{2}.$$

Since the above function of $\alpha$ is increasing, and $\alpha < 1$, we let $\alpha \to 1$ in the previous equation and define our quantities to be uniform in $\alpha$ as follows:

$$\widetilde{C}_1(\bar{x})\bar{x} = \exp(2\bar{x}) - 1 \;=\; \frac{1}{2},$$

that gives

$$\Delta x \cdot \|k\|_\infty \;=\; \bar{x} \;=\; \frac{1}{2}\log\left(\frac{3}{2}\right), \qquad \widetilde{C}_1(\bar{x}) \;=\; \frac{1}{\log\left(\frac{3}{2}\right)}.$$

Recalling the above definition of $\gamma$, we conclude that $\gamma \leq 1/4$ and therefore

$$C_1 \;=\; \frac{4}{3}\widetilde{C}_1(\bar{x}) \;=\; \frac{4}{3\log\left(\frac{3}{2}\right)}.$$

Finally, call $L$ the distance separating both the zero-waves $q_0$ and $\sigma_0$: one can pass to the limit $L \to 0$. By compactness, it converges to a non-interacting Riemann fan endowed with a zero-wave of size $q_0 + \sigma_0$. The size of the reflected wave reads:

$$|q_{-1}^{+}| = \sum_{n=1}^{\infty} |\sigma_{2n}^{+}| \leq C_1 \, |q_1^{-}| \cdot |\sigma_0|,$$

and the estimate (4.50) follows after taking (4.53) into account.

*Remark 4.3* There exists more direct manners to establish a quadratic estimate for the interaction of approaching waves like in Proposition 4.2. Indeed, consider for instance the proof of (4.49): one may proceed by just recalling the definition of sizes (4.19), so $\sigma_{-1} = J_m - J_r = f_r^{-} - f_m^{-}$ and the second identity in (4.51) becomes

$$q_1^{+} - q_1^{-} = (J_*^{-} - J_*^{+}) - \sigma_{-1}.$$

Recalling the definition of $\tilde{J}$, see (4.27), the quantities $J_*^{+}, J_*^{-}$ are given by

$$J_*^{+} = \tilde{J}(q_0, f_\ell^{+}, f_r^{-}) = \tilde{J}(q_0, f_\ell^{+}, f_m^{-} + \sigma_{-1}), \qquad J_*^{-} = \tilde{J}(q_0, f_\ell^{+}, f_m^{-}),$$

therefore, by the mean-value theorem, one derives:

$$J_*^{-} - J_*^{+} = -\frac{\partial \tilde{J}}{\partial (f^{-})}(q_0, f_\ell^{+}, f_m^{-} + \theta\sigma_{-1})\sigma_{-1}, \qquad \theta \in (0,1).$$

Notice that for $q_0 = 0$ one has $\tilde{J}(0, f^{\pm}) = f^{+} - f^{-}$, so we substitute

$$\frac{\partial \tilde{J}}{\partial (f^{-})}(0, f^{\pm}) \equiv -1, \qquad \text{for all } f^{-},$$

into the former expression. Accordingly we obtain:

$$q_1^{+} - q_1^{-} = -\sigma_{-1}\left[\frac{\partial \tilde{J}}{\partial (f^{-})}(q_0, f_\ell^{+}, f_m^{-} + \theta\sigma_{-1}) - \frac{\partial \tilde{J}}{\partial (f^{-})}(0, f_\ell^{+}, f_m^{-} + \theta\sigma_{-1})\right],$$

which finally furnishes,

$$|q_1^{+} - q_1^{-}| \leq |\sigma_{-1}| \, |q_0| \cdot \sup\left|\frac{\partial^2 \tilde{J}}{\partial \delta \, \partial (f^{-})}(\delta, f^{\pm})\right|.$$

The main issue is that resulting interaction constants cannot be easily expressed.

The following Proposition establishes a fundamental decay property:

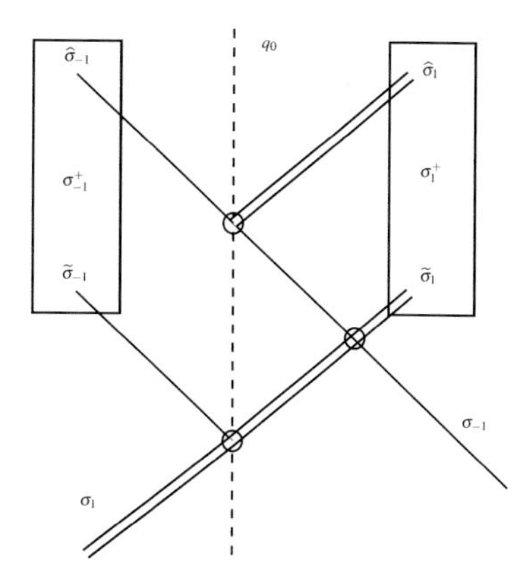


**Fig. 4.7** Schematic representation of the triple interaction

**Proposition 4.3** (Multiple interaction) *Assume that a* 1*-wave, a* 0*-wave and a* $-1$*-wave interact. Let* $\sigma^-_{-1}$, $\sigma^-_1$ *be the sizes of the incoming waves and* $\sigma^+_{-1}$, $\sigma^+_1$ *be the ones of the outgoing waves. Then*

$$|\sigma^+_{-1}| + |\sigma^+_1| \le |\sigma^-_{-1}| + |\sigma^-_1|. \tag{4.54}$$

*Besides, for* $\delta = a_r - a_\ell$*, one has*

$$\begin{cases} |\sigma^+_{-1}| - |\sigma^-_{-1}| & \le C_1\delta\left(|\sigma^-_{-1}| + |\sigma^-_1|\right) \\ |\sigma^+_1| - |\sigma^-_1| & \le C_1\delta\left(|\sigma^-_{-1}| + |\sigma^-_1|\right). \end{cases} \tag{4.55}$$

*Proof* Let interactions occur two at a time, to collect the result: see Fig. 4.7. The same procedure was previously used in [1].

- The first step is identical to the situation described in Lemma 4.2. Accordingly, the conclusion (4.40) holds for the present case, too. After the former interaction occurred, the wave of size $\widetilde{\sigma}_1$ will cross the $(-1)$-wave of size $\sigma^-_{-1}$ without changing size by linearity. The interaction between this last wave and the 0-wave produces two new waves, $\widehat{\sigma}_{\pm 1}$. Analogously, they satisfy

$$|\widehat{\sigma}_1| + |\widehat{\sigma}_{-1}| = |\sigma^-_{-1}|. \tag{4.56}$$

- Due to the linearity of $\pm 1$-waves, no other interaction can occur. The sizes of the outgoing waves $\sigma^+_{-1}$, $\sigma^+_1$ must satisfy

$$\sigma^+_{-1} = \widetilde{\sigma}_{-1} + \widehat{\sigma}_{-1}, \qquad \sigma^+_1 = \widetilde{\sigma}_1 + \widehat{\sigma}_1.$$

Therefore, collecting (4.40) and (4.56), we finally get (4.54):

$$|\sigma_{-1}^{+}| + |\sigma_{1}^{+}| \leq |\widetilde{\sigma}_{-1}| + |\widehat{\sigma}_{-1}| + |\widetilde{\sigma}_{1}| + |\widehat{\sigma}_{1}| \\ = |\sigma_{-1}^{-}| + |\sigma_{1}^{-}|.$$

- Finally let us prove (4.55) for the 1-family, the other one being analogous. From the construction above and Proposition 4.2, it is easy to deduce that

$$|\widetilde{\sigma}_1 - \sigma_1^{-}| \leq C_1 |\sigma_1^{-}| \delta, \qquad |\widehat{\sigma}_1| \leq C_1 |\sigma_{-1}^{-}| \delta.$$

  One has

$$|\sigma_1^{+}| - |\sigma_1^{-}| \leq |\widetilde{\sigma}_1| + |\widehat{\sigma}_1| - |\sigma_1^{-}| \ \leq \ |\widetilde{\sigma}_1 - \sigma_1^{-}| + |\widehat{\sigma}_1|,$$

  therefore we conclude thanks to the above estimates on $|\widetilde{\sigma}_1 - \sigma_1^{-}|$ and on $|\widehat{\sigma}_1|$.

## 4.5 An $L^1$ Error Estimate Through a Lyapunov Functional

The main objective is to quantify the gap between any two WB-approximations obtained with different grid parameters $(\Delta x)_1, (\Delta x)_2$. Two approximations $f_1^{\pm}, b_1(x)$ and $f_2^{\pm}, b_2(x)$ being given, at each point $(t, x)$, one considers the "transversal Riemann problem" for (4.14) with left/right data:

$$f_1^{\pm}(t, x), b_1(x), \qquad f_2^{\pm}(t, x), b_2(x).$$

We assume that $b_1, b_2$ are piecewise constant, non-decreasing, with jumps located in $(\Delta x)_1\mathbb{Z}$ and $(\Delta x)_2\mathbb{Z}$ respectively, and that they satisfy

$$\text{TV } b_1 \leq \text{TV } a, \qquad \text{TV } b_2 \leq \text{TV } a.$$

On the approximate initial data, we assume that

$$\text{TV} f_i^{-}(0, \cdot) + \text{TV} f_i^{+}(0, \cdot) \leq \text{TV } (f_0^{+}) + \text{TV } (f_0^{-}), \qquad i = 1, 2.$$

Let

$$q_{\pm 1}(t, x), \qquad q_0(x) = b_2(x) - b_1(x)$$

stand for the corresponding "transversal wave-strengths", and consider, for instance, that $f_1^{-}$ has a jump of size $\sigma$ at the point $(t, x_\alpha)$: see Fig. 4.8. In order to correctly devise the weights involved, it is necessary to know how the "transversal wave-strengths" evolve according to all the jumps present in both $f_1^{\pm}, b_1(x)$ and $f_2^{\pm}, b_2(x)$.

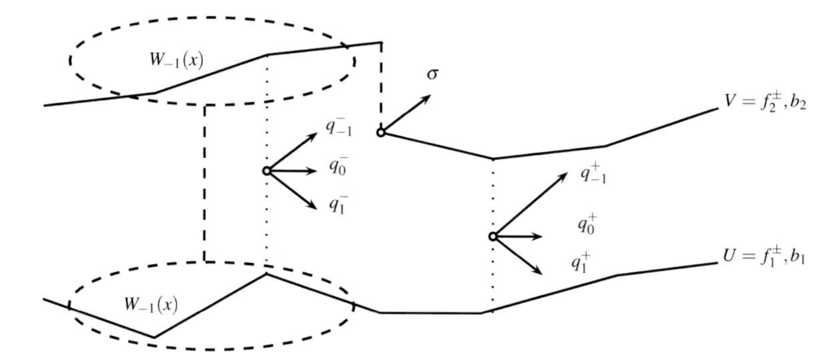


**Fig. 4.8** Interaction between a "transversal Riemann problem" (*left*) and a $-1$-wave resulting in a new Riemann problem (*right*) illustrating the simple situation described by Proposition 4.2, Case (c)

Hereafter, we use standard notations by Bressan [10]; the only exception is that the characteristic families are numbered $-1, 0, 1$ for obvious reasons. Let $U, V$ stand for $(f_1^-, f_1^+, b_1)$ and $(f_2^-, f_2^+, b_2)$ respectively. We write $\sigma_i^\alpha$ for the size a front located at $x^\alpha$, of the family $i \in \{-1, 0, 1\}$; zero-waves are measured simply by the jump of $b_1(x)$ or $b_2(x)$, respectively for $U$ or $V$. All $\sigma_0^\alpha$ are positive, because $b_1(x)$ and $b_2(x)$ are monotone, non-decreasing.

The Lyapunov functional $\Phi[U, V]$ reads, for $x_1 < x_2$ and $t \leq T = (x_2 - x_1)/2$:

$$\boxed{t \mapsto \Phi[U, V](t) = \int_{x_1+t}^{x_2-t} |q_0(x)| W_0(t, x) dx + \sum_{i=\pm 1} \int_{x_1+t}^{x_2-t} |q_i(t, x)| W_i(x) dx,} \tag{4.57}$$

where the weights $W_i$ are defined as follows:

$$\begin{aligned} W_0(t, x) &= 1 + \kappa_1 A_0(t, x) + \kappa_2 \big(Q(U) + Q(V)\big), \\ W_i(x) &= 1 + \kappa_1 A_i(x), \qquad i = -1,\ 1 \end{aligned}$$

and

$$\begin{aligned} A_0(t, x) &= \sum_{x_\alpha < x} |\sigma_1^\alpha| + \sum_{x_\alpha > x} |\sigma_{-1}^\alpha|, \\ A_{-1}(x) &= \sum_{x_\alpha < x} \sigma_0^\alpha, \quad [\text{ 0-fronts on the left of } x], \\ A_1(x) &= \sum_{x_\alpha > x} \sigma_0^\alpha, \quad [\text{ 0-fronts on the right of } x]. \end{aligned}$$

The sums above extend over all jumps in $U$ and $V$. An estimate for $A_{\pm 1}$ reads:

$$A_{\pm 1}(x) \;\le\; \mathrm{TV} b_1 + \mathrm{TV} b_2 \;\le\; 2\mathrm{TV}\, a.$$

On the other hand, an estimate on $A_0$ goes as follows. By defining

$$\mathscr{A}_0 \doteq \mathrm{TV} f_0^- + \mathrm{TV}\, f_0^+ + 2C_0\, \mathrm{TV}\, a, \tag{4.58}$$

and recalling (4.32), one obtains

$$\begin{aligned} A_0(t,x) &\le L_\pm(t;U) + L_\pm(t;V) \\ &\le \mathrm{TV}\, f_1^-(0,\cdot) + \mathrm{TV}\, f_1^+(0,\cdot) + \mathrm{TV}\, f_2^-(0,\cdot) + \mathrm{TV}\, f_2^+(0,\cdot) + 2C_0\,(\mathrm{TV}\, b_1 + \mathrm{TV}\, b_2) \\ &\le 2\mathscr{A}_0. \end{aligned}$$

As usual, $Q(U)$, $Q(V)$ stand for interaction potentials between $\pm 1$-waves and 0-waves showing up in $U$, $V$ respectively:

$$Q(U)(t) = \sum_\beta \sigma_0^\beta \left[ \sum_{\alpha,\, x_\alpha < x_\beta} |\sigma_1^\alpha| + \sum_{\alpha,\, x_\alpha > x_\beta} |\sigma_{-1}^\alpha| \right]$$

where the sum runs over all jumps of $U$ in $(x_1 + t, x_2 - t)$. Hence

$$Q(U)(t) \;\le\; \mathrm{TV}\,\{b_1\}\, L_\pm(t;U) \;\le\; \mathrm{TV}\,\{a\}\, L_\pm(0+,U) \;\le\; \mathrm{TV}\{a\}\, \mathscr{A}_0.$$

The situation is analogous for $V$. Therefore we estimate the sum of the $Q$ as follows:

$$Q(U) + Q(V) \le 2\mathrm{TV}\,\{a\}\, \mathscr{A}_0.$$

In order to control the size of these weights, one must manage the bounds:

$$W_{\pm 1}(x) \le 1 + 2\kappa_1 \mathrm{TV}\, a, \tag{4.59}$$

$$W_0(t,x) \le 1 + 2\mathscr{A}_0\, (\kappa_1 + \kappa_2 \mathrm{TV}\,\{a\})\,. \tag{4.60}$$

The constants $\kappa_1$, $\kappa_2$ still have to be determined. Here we are going to specialize the analysis presented in [10, 11] for more general systems and avoid the smallness conditions on the initial data. Let us present the main steps of the analysis:

1. Show that the functional decreases outside interaction times: see Lemma 4.3. A natural bound on TV $a$ follows and $\kappa_1$ is suitably chosen, see Remark 4.4.
2. Show that the functional decreases at interaction times: see Lemma 4.4. The constant $\kappa_2$ is chosen at this step.
3. Quantify the relation between $\Phi[U,V](t)$ and the $L^1$ difference between the two approximate solutions, done in Lemma 4.5.

The next two lemmas state that $t \mapsto \Phi[U, V](t)$ decreases both outside interaction times (Lemma 4.3) and at interaction times (Lemma 4.4).

### 4.5.1 Decay Outside Interactions

**Lemma 4.3** *Let $U(t, \cdot)$ and $V(t, \cdot)$ be two approximate solutions generated by the Well-Balanced algorithm, out of the initial data*

$$U_0 = \left(f_1^{\pm}(t = 0, \cdot), a(\cdot)\right), \qquad V_0 = \left(f_2^{\pm}(t = 0, \cdot), b(\cdot)\right).$$

*Let $K > 0$ such that the weights $W_{\pm 1}$ satisfy a uniform bound of the following type:*

$$1 \leq W_{\pm 1}(x) \leq K \tag{4.61}$$

*and assume that*

$$\kappa_1 \geq 2KC_1 \tag{4.62}$$

*with $C_1$ as in (4.48). Then, outside interaction times, one has*

$$\frac{d\Phi[U, V]}{dt} \leq 0.$$

*Remark 4.4* From (4.5), one can choose $K = 1 + 2\kappa_1 \mathrm{TV}\ a$ and rewrite (4.62) as

$$\kappa_1 \geq 2C_1 \left[1 + 2\kappa_1 \mathrm{TV}\ a\right].$$

The above inequality is possible whenever (see (4.48))

$$4C_1 \mathrm{TV}\ a < 1, \quad \Leftrightarrow \quad 16\mathrm{TV}\ a \leq 3\log(3/2). \tag{4.63}$$

Therefore, provided that (4.63) holds, we can operate the following choice:

$$\kappa_1 = \frac{2C_1}{1 - 4C_1 \mathrm{TV}\ a}, \qquad K = \frac{\kappa_1}{2C_1} = \frac{1}{1 - 4C_1 \mathrm{TV}\ a}. \tag{4.64}$$

*Proof* Now we prove Lemma 4.3. Following Bressan (see [10, p. 155]), outside interaction times it is convenient to write the time-derivative of $\Phi$ as follows:

$$\begin{aligned} \frac{d\Phi[U, V]}{dt} &= \sum_{i=-1}^{1} |q_i(x)| W_i(x)(-1 + \lambda_i)\Big|_{x = x_1 + t} \\ &\quad + \sum_{i=-1}^{1} |q_i(x)| W_i(x)(-1 - \lambda_i)\Big|_{x = x_2 - t} + \sum_{\alpha} \sum_{i=-1}^{1} E_{\alpha,i}, \end{aligned}$$

being

$$E_{\alpha,i} = |q_i^{\alpha+}| W_i^{\alpha+} (\lambda_i^{\alpha+} - \dot{x}^\alpha) - |q_i^{\alpha-}| W_i^{\alpha-} (\lambda_i^{\alpha-} - \dot{x}^\alpha)$$
$$= \left[|q_i^{\alpha+}| W_i^{\alpha+} - |q_i^{\alpha-}| W_i^{\alpha-}\right] (\lambda_i^\alpha - \dot{x}^\alpha)$$

where we used that the $\lambda_i$'s are constant. Since $|\lambda_i| \leq 1$, the contribution from the boundaries is non-positive and then:

$$\frac{d\Phi[U,V]}{dt} \leq \sum_\alpha \sum_{i=-1}^{1} E_{\alpha,i}.$$

Thanks to the linear structure of families $\pm 1$, lots of simplification occur in the sum above. For instance, if $i = k_\alpha$ then the corresponding speeds coincide, $\lambda_i^\alpha = \dot{x}^\alpha$, and thus $E_{\alpha,i} = 0$. We shall analyze the jumps that occur in the $V = (f_2^\pm, b_2)$ vector of unknowns; the analysis for the jumps in $U$ is completely similar (see also [10, p. 160]). Such a framework exactly meets with the interaction estimates given in Proposition 4.2. Accordingly, let $k_\alpha \in \{\pm 1, 0\}$ denote the characteristic family of the jump present at the abscissa $x_\alpha$. To carry on, one distinguishes between each value of $k_\alpha$. For simplicity, in the following we will often omit the dependence on $\alpha$.

- If $k_\alpha = -1 = \dot{x}^\alpha$, an easy computation shows that $E_{-1} = 0$ and that

  $$E_0 = |q_0^+| W_0^+ - |q_0^-| W_0^-, \qquad E_1 = 2\left[|q_1^+| W_1^+ - |q_1^-| W_1^-\right].$$

  Moreover we have

  $$q_0^+ = q_0^-, \qquad W_1^+ = W_1^-, \qquad W_0^+ - W_0^- = -\kappa_1 |\sigma_{-1}|$$

  and hence

  $$\sum_{i=-1}^{1} E_i = E_0 + E_1 = -\kappa_1 |\sigma_{-1}| |q_0^-| + 2\left\{|q_1^+| - |q_1^-|\right\} W_1^-.$$

  From (4.49), Case (a) of Proposition 4.2, it follows that $|q_1^+| \leq |q_1^-| + C_1 |q_0^-| |\sigma_{-1}|$. Also, recalling (4.61), the weight $W_1^-$ is supposed to be smaller than $K$, so

  $$\sum_{i=-1}^{1} E_i \leq |q_0^-| |\sigma_{-1}| (-\kappa_1 + 2KC_1) \leq 0.$$

- If $k_\alpha = 1 = \dot{x}^\alpha$, this is the simple Case (c), and

$$\sum_{i=-1}^{1} E_i = E_{-1} + E_0$$
$$= -2\left\{|q_{-1}^+|W_{-1}^+ - |q_{-1}^-|W_{-1}^-\right\} \;-\; \left\{|q_0^+|W_0^+ - |q_0^-|W_0^-\right\}.$$

Here $q_0, q_{-1}, W_{-1}$ do not change, while

$$W_0^+ - W_0^- = +\kappa_1\sigma_1.$$

Hence one gets a negative $\kappa_1 > 0$:

$$\sum_{i=-1}^{1} E_i = -|q_0|\left\{W_0^+ - W_0^-\right\} = -\kappa_1|q_0|\sigma_1 \le 0.$$

- If $k_\alpha = 0 = \dot{x}^\alpha$, this is Case (b), depicted in Fig. 4.6, with $\dot{x} = \lambda_0 = 0$ so $E_0 = 0$.

$$\sum_{i=-1}^{1} E_i = E_{-1} + E_1$$
$$= -\left\{|q_{-1}^+|W_{-1}^+ - |q_{-1}^-|W_{-1}^-\right\} \;+\; \left\{|q_1^+|W_1^+ - |q_1^-|W_1^-\right\}.$$

The weights $W_i^\pm$, $i = \pm 1$ jump as follows:

$$W_{-1}^+ - W_{-1}^- = +\kappa_1|\sigma_0| \ge 0, \qquad W_1^+ - W_1^- = -\kappa_1|\sigma_0|.$$

Hence, by means of (4.50), we find that

$$\begin{aligned} E_{-1} &= -|q_{-1}^+|\left\{W_{-1}^+ - W_{-1}^-\right\} - W_{-1}^-\left\{|q_{-1}^+| - |q_{-1}^-|\right\} \\ &\le -W_{-1}^-\left\{|q_{-1}^+| - |q_{-1}^-|\right\} \\ &\le K|q_{-1}^- - q_{-1}^+| \;\le\; KC_1\,\sigma_0\,|q_1^-| \end{aligned}$$

while, in a quite similar way,

$$\begin{aligned} E_1 &= |q_1^-|(W_1^+ - W_1^-) + (|q_1^+| - |q_1^-|)W_1^+ \\ &\le -\kappa_1\sigma_0|q_1^-| + K\left|q_1^+ - q_1^-\right| \\ &\le -\kappa_1\sigma_0|q_1^-| + KC_1\sigma_0|q_1^-| \\ &\le \sigma_0|q_1^-|(KC_1 - \kappa_1). \end{aligned}$$

At this point, having $\kappa_1 \ge 2KC_1$ again ensures $E_{-1} + E_1 \le 0$.

### 4.5.2 Decay at Interaction Times

**Lemma 4.4** *In the assumptions of Lemma 4.3, assume that (4.63) holds and that*

$$\kappa_2 \geq \frac{\kappa_1 C_1}{1 - C_1 \mathrm{TV}\, a}. \tag{4.65}$$

*Then $\Phi[U, V](t)$ decreases at interaction times.*

*Proof* Assume that at a certain time $t$ interactions occur for the approximate solution $U$. Recalling the definition (4.57) of $\Phi$, we notice that the $|q_{\pm 1}(t, x)|$ change continuously in $L^1_{loc}$. The only term that can change in a discontinuous way across the interaction time $t$ is the weight $W_0(t, x)$:

$$\Delta W_0(t, x) = \kappa_1 \Delta A_0(t, x) + \kappa_2 \Delta Q(U)(t)$$

The term $\Delta A_0$ can increase across interaction times, while $\Delta Q(U)(t)$ decreases, as follows. For each $x_\beta$ where a 0-wave is located, let

$$\sigma_0^\beta, \qquad \sigma_{-1}^{\beta\pm}, \qquad \sigma_1^{\beta\pm}$$

the waves involved in the interaction (with obvious notation). Thanks to Proposition 4.3, one of the two terms

$$|\sigma_1^{\beta+}| - |\sigma_1^{\beta-}|, \qquad |\sigma_{-1}^{\beta+}| - |\sigma_{-1}^{\beta-}|$$

is negative, while the other one is bounded by $C_1 \sigma_0^\beta \left( |\sigma_1^{\beta-}| + |\sigma_{-1}^{\beta-}| \right)$. Hence,

$$\begin{aligned}
\Delta Q = & -\sum_\beta \sigma_0^\beta \left( |\sigma_1^{\beta-}| + |\sigma_{-1}^{\beta-}| \right) \\
& + \sum_\beta \left( |\sigma_1^{\beta+}| - |\sigma_1^{\beta-}| \right) \mathrm{TV}\{a; (x_\beta, \infty)\} + \sum_\beta \left( |\sigma_{-1}^{\beta+}| - |\sigma_{-1}^{\beta-}| \right) \mathrm{TV}\{a; (-\infty, x_\beta)\} \\
\leq & \, (-1 + C_1 \mathrm{TV}\, a) \sum_\beta \sigma_0^\beta \left( |\sigma_1^{\beta-}| + |\sigma_{-1}^{\beta-}| \right).
\end{aligned}$$

On the other hand, thanks to (4.55) in Proposition 4.3, the possible increase of $A_0$ is bounded uniformly in $x$ as follows:

$$\Delta A_0(t, x) \leq C_1 \sum_\beta \sigma_0^\beta \left( |\sigma_1^{\beta-}| + |\sigma_{-1}^{\beta-}| \right).$$

Therefore

$$\Delta W_0 \leq (\kappa_1 C_1 - \kappa_2 (1 - C_1 \mathrm{TV}\, a)) \sum_\beta \sigma_0^\beta \left( |\sigma_1^{\beta-}| + |\sigma_{-1}^{\beta-}| \right),$$

which is $\leq 0$ when $1 - C_1 \mathrm{TV}\, a > 0$, ensured by (4.63), and when $\kappa_2$ satisfies (4.65).

*Remark 4.5* Following Remark 4.4, here we summarize the choice of $\kappa_1, \kappa_2$ and the bounds on $W_i$ obtained so far. Thanks to Lemma 4.3, we have

$$W_{\pm 1} \leq 1 + 2\kappa_1 \mathrm{TV} a \; \leq \; K = \frac{\kappa_1}{2C_1};$$

this is possible if (4.63) holds, that is $4C_1 \mathrm{TV}\, a < 1$. Then $\kappa_1$ can be set as (4.64). Therefore a bound for $W_{\pm 1}$ in terms of the data is:

$$W_{\pm 1} \leq \frac{1}{1 - 4C_1 \mathrm{TV} a} = K. \tag{4.66}$$

Recalling (4.59) and thanks to Lemma 4.4, we get

$$\begin{aligned} W_0(t,x) &\leq 1 + 2\mathscr{A}_0 (\kappa_1 + \kappa_2 \mathrm{TV}\, \{a\}) \; \leq \; 1 + 2\frac{\kappa_1 \mathscr{A}_0}{1 - C_1 \mathrm{TV}\, \{a\}} \\ &= 1 + \frac{4C_1 \mathscr{A}_0}{(1 - 4C_1 \mathrm{TV}\, a)(1 - C_1 \mathrm{TV}\, \{a\})} \doteq K_0. \end{aligned} \tag{4.67}$$

### 4.5.3 Decay of Lyapunov Functional for Weak Relaxation

**Lemma 4.5** *By equivalence of $\Phi$ with the $L^1$ distance between two approximations,*

$$I(t) \; = \; \int_{x_1+t}^{x_2-t} |f_1^+(t,x) - f_2^+(t,x)| + |f_1^-(t,x) - f_2^-(t,x)| dx$$

*satisfies the estimate*

$$\boxed{I(t) \leq K \cdot I(0) + (2KC_0 + K_0) \int_{x_1}^{x_2} |b_1 - b_2| \, dx + (2C_0 - 1) \int_{x_1+t}^{x_2-t} |b_1 - b_2| \, dx.} \tag{4.68}$$

*Remark 4.6* According to (4.67), $K_0 = 1 + K \cdot \frac{4C_1}{1-C_1 \mathrm{TV}\, a}\mathscr{A}_0$, but simultaneously,

$$K = \frac{1}{1 - 4C_1 \mathrm{TV}\, a}, \qquad 1 - C_1 \mathrm{TV}\, a = \frac{3K+1}{4K}, \qquad C_1 \le \frac{14}{3}.$$

So, $K_0 = 1 + \frac{16\, K^2 C_1}{3K+1}\mathscr{A}_0$ and for instance, if $-x_1, x_2 \to +\infty$, then (4.68) rewrites

$$I_{\mathbb{R}}(t) \le K \cdot I_{\mathbb{R}}(0) + [2C_0(K+1) + K_0 - 1] \int_{\mathbb{R}} |b_1 - b_2|\, dx$$

and therefore

$$\boxed{I_{\mathbb{R}}(t) \le K \cdot I_{\mathbb{R}}(0) \;+\; 2[C_0(K+1) + \frac{8K^2 C_1}{3K+1}\mathscr{A}_0] \int_{\mathbb{R}} |b_1 - b_2|\, dx.} \tag{4.69}$$

Notice also that the quantity $(2C_0 - 1)$ in (4.68) can be negative.

*Proof* Recalling (4.24) and using $W_{\pm 1} \ge 1$, one gets

$$\begin{aligned} I(t) &\le \int_{x_1+t}^{x_2-t} \{|q_1| + |q_{-1}| + |b_1 - b_2|\}\, dx \;+\; (2C_0 - 1) \int_{x_1+t}^{x_2-t} |b_1 - b_2|\, dx \\ &\le \Phi[U,V](t) + (2C_0 - 1) \int_{x_1+t}^{x_2-t} |b_1 - b_2|\, dx \end{aligned}$$

and also, always taking advantage of (4.24),

$$\begin{aligned} \Phi[U,V](t) &\le K \sum_{i=-1,1} \int_{x_1+t}^{x_2-t} |q_i| dx \;+\; K_0 \int_{x_1+t}^{x_2-t} |b_1 - b_2| dx \\ &\le KI(t) \;+\; (2KC_0 + K_0) \int_{x_1+t}^{x_2-t} |b_1 - b_2|\, dx. \end{aligned}$$

Altogether, since $t \mapsto \Phi[U,V](t)$ decreases, it comes that:

$$\begin{aligned} I(t) &\le \Phi[U,V](t) + (2C_0 - 1) \int_{x_1+t}^{x_2-t} |b_1 - b_2|\, dx \\ &\le \Phi[U,V](0) + (2C_0 - 1) \int_{x_1+t}^{x_2-t} |b_1 - b_2|\, dx \\ &\le KI(0) + (2KC_0 + K_0) \int_{x_1}^{x_2} |b_1 - b_2|\, dx + (2C_0 - 1) \int_{x_1+t}^{x_2-t} |b_1 - b_2|\, dx \end{aligned}$$

which is precisely (4.68).

With the time-decay of $\Phi[U, V](t)$ at hand, by just selecting $\Delta x = (\Delta x)_1$,

$$b_1 = P^{(\Delta x)}a, \qquad \partial_x a(x) = k(x), \tag{4.70}$$

$V(t = 0, \cdot) = P^{(\Delta x}U(t = 0, \cdot)$ and sending $(\Delta x)_2 \to 0$, one obtains that the global $L^1$ error of the WB scheme at any time $t > 0$ is bounded by

$$\int_{x_1+t}^{x_2-t} |f^{\pm}_{\Delta x}(t,x) - f^{\pm}(t,x)|dx \le K \int_{x_1}^{x_2} |f^{\pm}_{\Delta x}(0,x) - f^{\pm}(0,x)|dx$$
$$+ \big(2KC_0 + K_0\big)\Delta x\, \mathrm{TV}\,\{a; (x_1, x_2)\} + \big(2C_0 - 1\big)\Delta x\, \mathrm{TV}\,\{a; (x_1 + t, x_2 - t)\},$$

where $K$, $K_0$ are given by (4.66), (4.67) respectively. By (4.69), the global $L^1$ error on the whole real line is bounded uniformly in time by the quantity,

$$\frac{1}{\Delta x}\int_{\mathbb{R}} |f^{\pm}_{\Delta x}(t,x) - f^{\pm}(t,x)|dx \tag{4.71}$$

$$\le K\mathrm{TV}(f^{\pm}(0,\cdot)) + 2\left[C_0(K+1) + \frac{8K^2C_1}{3K+1}\mathscr{A}_0\right]\|k\|_{L^1(\mathbb{R})} \tag{4.72}$$

where $\mathscr{A}_0$ is defined at (4.58), which blows up as

$$\|k\|_{L^1(\mathbb{R})} \to \frac{1}{4C_1} = \frac{3}{16}\log(4/3) \doteq C \tag{4.73}$$

since the constant $K$ does (see (4.64) and recall (4.48)). This was to be expected, as for stiff relaxation regimes and well-prepared initial data, one expects $\rho = f^+ + f^-$ to match the entropy solution of the conservation law $\partial_t\rho + \partial_x A(\rho) = 0$, and one cannot have order 1 convergence as $\Delta x \to 0$. This completes the proof of the first estimate, $\mathscr{E}_1$, in Theorem 4.1. The second estimate, relying on Kuznetsov's method and entropy dissipation inequalities, is forwarded in the next Chap. 5.

*Remark 4.7* By using results in Sect. 4.5, the global $L^1$ error estimate for a locally damped semi-linear wave equation (obtained in [4]) can be refined: consider,

$$\partial_t\rho + \partial_x J = 0 \quad \partial_t J + \partial_x\rho = -2k(x)g(J) \tag{4.74}$$

with $g \in C^1(\mathbb{R})$, $g(0) = 0$, $g$ strictly increasing. Recalling the assumptions (4.6) and (4.5), we are now considering a case where $A = 0$ but with a more general dependence on $J$, being $g(J)$ possibly nonlinear. Propositions 4.2 and 4.3 are still valid (see the corresponding Propositions in [4], with $C_1 = Lip(g)$. Moreover, smallness restrictions on $\Delta x$ as in (4.48) drop. Indeed, if both $A = 0$ and $\alpha = 0$ in Lemma 4.1, the constant $\widetilde{C}_1$ given in (4.42) can be set to unity regardless of the size of $\delta = a_r - a_\ell$. Differently, when $\alpha > 0$, the size of $\widetilde{C}_1$ can grow exponentially with $\delta$; this was the reason for devising a bound on $\delta$, and correspondingly on $\Delta x$ (assuming that $k$ is locally bounded). In conclusion, for the system (4.74) the same estimates (4.68), (4.69) and (4.71) hold with the following values for the constants:

$$C_0 = \|g\|_\infty, \qquad C_1 = Lip(g).$$

There are no restrictions on both $\mathrm{TV} f_0^{\pm}$ and $\Delta x \geq 0$, beyond the condition,

$$K = (1 - 4C_1 \mathrm{TV}\ a)^{-1} < \infty, \quad \text{that is,} \quad \mathrm{TV}\ a = \|k\|_{L^1} < \frac{1}{4\,Lip(g)}.$$

## References


1. D. Amadori, A. Corli, Glimm estimates for a model of multiphase flow. Preprint (2012)
2. D. Amadori, L. Gosse, Transient $L^1$ error estimates for well-balanced schemes on non-resonant scalar balance laws. J. Differ. Equ. **255**, 469–502 (2013)
3. D. Amadori, L. Gosse, Stringent error estimates for one-dimensional, space-dependent $2 \times 2$ relaxation systems. Ann. Inst. H. Poincaré Anal. Nonlinéaire (2015), http://dx.doi.org/10.1016/j.anihpc.2015.01.001
4. D. Amadori, L. Gosse, Error estimates for well-balanced and time-split schemes on a locally damped semilinear wave equation. Math. Comput. http://dx.doi.org/10.1090/mcom/3043
5. D. Amadori, G. Guerra, Global BV solutions and relaxation limit for a system of conservation laws. Proc. R. Soc. Edinb. Sect. A **131**, 1–26 (2001)
6. S. Bianchini, A Glimm-type functional for a special Jin-Xin relaxation model. Ann. Inst. H. Poincaré Anal. Non Linéaire **18**(1), 19–42 (2001)
7. S. Bianchini, Relaxation limit of the Jin-Xin relaxation model. Commun. Pure Appl. Math. **59**(5), 688–753 (2006)
8. S. Bianchini, B. Hanouzet, R. Natalini, Asymptotic behavior of smooth solutions for partially dissipative hyperbolic systems with a convex entropy. Commun. Pure Appl. Math. **60**, 1559–1622 (2007)
9. F. Bouchut, B. Perthame, Kružkov's estimates for scalar conservation laws revisited. Trans. Am. Math. Soc. **350**(7), 2847–2870 (1998)
10. A. Bressan, *Hyperbolic Systems of Conservation Laws. The One-dimensional Cauchy problem*. Oxford Lecture Series in Mathematics and its Applications, vol. 20 (Oxford University Press, Oxford, 2000)
11. A. Bressan, T.-P. Liu, T. Yang, $L^1$ stability estimates for $n \times n$ conservation laws. Arch. Ration. Mech. Anal. **149**, 1–22 (1999)
12. F. Coquel, S. Jin, J.-G. Liu, L. Wang, Well-posedness and singular limit of a semilinear hyperbolic relaxation system with a two-scale discontinuous relaxation rate. Arch. Ration. Mech. Anal. **214**, 1051–1084 (2014)
13. G.A.M. de Almeida, P. Bates, Applicability of the local inertial approximation of the shallow water equations to flood modelling. Water Resour. Res. **49**, 4833–4844 (2013)
14. A. Edwards, B. Perthame, N. Seguin, M. Tournus, Analysis of a simplified model of the urine concentration mechanism. Netw. Heterog. Media **7**, 989–1018 (2012)
15. J. Glimm, D.H. Sharp, An $S$-matrix theory for classical nonlinear physics. Found. Phys. **16**, 125–141 (1986)
16. A. Gnudi, F. Odeh, M. Rudan, Investigation of non-local transport phenomena in small semiconductor devices. Trans. Emerg. Telecomun. Technol. **1**, 307–312 (1990)
17. L. Gosse, Time-splitting schemes and measure source terms for a quasilinear relaxing system. M3AS **13**(8), 1081–1101 (2003)

18. L. Gosse, *Computing Qualitatively Correct Approximations of Balance Laws*. SIMAI Springer Series, vol. 2 (Springer, Berlin, 2013)
19. L. Gosse, Redheffer products and numerical approximation of currents in one-dimensional semiconductor kinetic models. SIAM Multiscale Model. Simul. **12**(4), 1533–1560 (2014)
20. L. Gosse, G. Toscani, Space localization and well-balanced schemes for discrete kinetic models in diffusive regimes. SIAM J. Numer. Anal. **41**, 641–658 (2004)
21. L. Gosse, A. Tzavaras, Convergence of relaxation schemes to the equations of elastodynamics. Math. Comput. **70**(234), 555–577 (2001)
22. J.W. Jerome, C.-W. Shu, Transport effects and characteristic modes in the modeling and simulation of submicron devices. IEEE Trans. Comput.-Aided Des. Integr. Circuits Syst. **14**, 917–923 (1995)
23. S. Jin, Z. Xin, The relaxation schemes for systems of conservation laws in arbitrary space dimension. Commun. Pure Appl. Math. **48**, 235–276 (1995)
24. M. Laforest, A posteriori error estimate for front-tracking: system of conservation laws. SIAM J. Math. Anal. **35**, 1347–1370 (2004)
25. H. Liu, R. Natalini, Long-time diffusive behavior of solutions to a hyperbolic relaxation system. Asymptot. Anal. **25**, 21–38 (2001)
26. H.L. Liu, G. Warnecke, Convergence rates for relaxation schemes approximating conservation laws. SIAM J. Numer. Anal. **37**(4), 1316–1337 (2000)
27. C. Mascia, K. Zumbrun, Pointwise Green's function bounds and stability of relaxation shocks. Indiana Univ. Math. J. **51**(4), 773–904 (2001)
28. R. Natalini, Recent results on hyperbolic relaxation problems, *Analysis of Systems of Conservation Laws (Aachen, 1997)*. Monographs and Surveys in Pure and Applied Mathematics, vol. 99 (Chapman & Hall/CRC, Boca Raton, 1999)
29. B. Perthame, N. Seguin, M. Tournus, A simple derivation of BV bounds for inhomogeneous relaxation systems. Commun. Math. Sci. **13**, 577–586 (2015)
30. J. Rauch, M. Reed, Jump discontinuities of semilinear, strictly hyperbolic systems in two variables: creation and propagation. Commun. Math. Phys. **81**, 203–227 (1981)
31. F. Sabac, The optimal convergence rate of monotone finite difference methods for hyperbolic conservation laws. SIAM J. Numer. Anal. **34**, 2306–2318 (1997)
32. D. Serre, Relaxation semi-linéaire et cinétique des systèmes de lois de conservation. Ann. Inst. Henri Poincaré Anal. Non Linéaire **17**(2), 169–192 (2000)
33. M.A. Stettler, M.A. Alam, M.S. Lundstrom, A critical examination of the assumptions underlying macroscopic transport equations for silicon devices. IEEE Trans. Electron Dev. **10**, 733–740 (1993)
34. E. Tadmor, T. Tang, Pointwise error estimates for relaxation approximations to conservation laws. SIAM J. Math. Anal. **32**(4), 870–886 (2000)
35. L.R. Tcheugoué Tébou, E. Zuazua, Uniform exponential long time decay for the space semi-discretization of a locally damped wave equation via an artificial numerical viscosity. Numer. Math. **95**, 563–598 (2003)
36. Z.H. Teng, First-order $L^1$ convergence for relaxation approximations to conservation laws. Commun. Pure Appl. Math. **51**(8), 857–895 (1998)
37. E. Weinan, Homogenization of scalar conservation laws with oscillatory forcing terms. SIAM J. Appl. Math **52**, 959–972 (1992)

# Chapter 5
# Entropy Dissipation and Comparison with Lyapunov Estimates

**Abstract** In this chapter we analyze the scheme, which was introduced in the previous chapter, by means of a classical Kuznetsov approach. An alternative qualitative estimate, in terms of time and mesh size, is therefore devised. The two estimates are compared, revealing complementary aspects.

**Keywords** Well-balanced schemes · Accuracy of schemes for balance laws · Kuznetsov approach

The estimate proved in Chap. 4 suits well the non-stiff case for (4.1). However, one may feel the need for a study of the complementary situation, where typically $|k(x)|\Delta x$ can become (locally) big. In order to quantify the $L^1$ error of WB schemes in this context too, we adapt a method of [8] (see also [6]) based on entropy dissipation and inspired by the seminal ideas of Kuznetsov [7].

## 5.1 A Time-Dependent $L^1$ Error Estimate $\mathscr{E}_2$

### *5.1.1 Entropy Dissipation Inequalities*

Let us first describe what type of entropy inequalities are satisfied by the exact solution and by the WB approximation. On one hand, the exact solution of (4.8) is such that, for any constant values $k_\pm \in \mathbb{R}$ and any test-function $0 \le \varphi(t,x) \in C_0^\infty(\mathbb{R}^+ \times \mathbb{R})$,

$$
\begin{aligned}
&-\int_0^T \int_{\mathbb{R}} \left(|f^+ - k_+| + |f^- - k_-|\right) \partial_t \varphi + \left(|f^+ - k_+| - |f^- - k_-|\right) \partial_x \varphi \cdot dx \cdot dt \\
&+ \int_{\mathbb{R}} \left(|f^+(T,x) - k_+| + |f^-(T,x) - k_-|\right) \varphi(T,x) \cdot dx \\
&- \int_{\mathbb{R}} \left(|f^+(0,x) - k_+| + |f^-(0,x) - k_-|\right) \varphi(0,x) \cdot dx
\end{aligned}
$$



D. Amadori and L. Gosse, *Error Estimates for Well-Balanced Schemes on Simple Balance Laws*, SpringerBriefs in Mathematics,
DOI 10.1007/978-3-319-24785-4_5

$$\leq \int_0^T \int_{\mathbb{R}} k(x) \left( sgn(f^+ - k_+) - sgn(f^- - k_-) \right) G(f^+, f^-)\varphi \cdot dx \cdot dt. \tag{5.1}$$

On the other hand, the WB approximation is the exact solution of (4.14) with piecewise-constant initial data fitted to the length separating 2 zero-waves (see again Fig. 4.3), in particular there is no projection at each time-step.

**Lemma 5.1** *For any test-function $\varphi(t, x) \geq 0$ compactly supported on $(0, T) \times \mathbb{R}$,*

$$\begin{aligned}
&- \int_0^T \int_{\mathbb{R}} \left( |f^+ - k_+| + |f^- - k_-| \right) \partial_t \varphi + \left( |f^+ - k_+| - |f^- - k_-| \right) \partial_x \varphi \cdot dx \cdot dt \\
&- \int_0^T \int_{\mathbb{R}} k(x) \left( sgn(f^+ - k_+) - sgn(f^- - k_-) \right) G(f^+, f^-)\varphi \cdot dx \cdot dt \\
&\leq C_\alpha \sum_{n,j} \mathrm{TV}\left( f^\pm(t^n, \cdot); \{x_{j-1}, x_j\} \right) \int_{t^n}^{t^{n+1}} \int_{x_{j-1}}^{x_j} k(x)\varphi(t, x_{j-\frac{1}{2}})\, dx \cdot dt \\
&+ C_\beta \sum_{n,j} \int_{t^n}^{t^{n+1}} \int_{x_{j-1}}^{x_j} k(x) \left| \varphi(t, x) - \varphi(t, x_{j-\frac{1}{2}}) \right| dx \cdot dt,
\end{aligned} \tag{5.2}$$

*where $C_\alpha = Lip(G)$ and $C_\beta = 2C_0$, the Maxwellian gap defined in (4.25).*

*Proof* The proof is divided into several steps.

- Using the standard notation, $t^n = n\Delta t$, $C_j = (x_{j-\frac{1}{2}}, x_{j+\frac{1}{2}})$ with $x_{j-\frac{1}{2}} = (j - \frac{1}{2})\Delta x$ the locus of the zero-waves, comes in each "cell" $C_j \times (t^n, t^{n+1})$,

$$\begin{aligned}
&- \int_{t^n}^{t^{n+1}} \int_{C_j} \left( \eta_+(f^+) + \eta_-(f^-) \right) \partial_t \varphi + \left( \eta_+(f^+) - \eta_-(f^-) \right) \partial_x \varphi \cdot dx \cdot dt \\
&+ \int_{C_j} \left( \eta_+(f^+(t^{n+1}, x)) + \eta_-(f^-(t^{n+1}, x)) \right) \varphi(t^{n+1}, x) \cdot dx \\
&- \int_{C_j} \left( \eta_+(f^+(t^n, x)) + \eta_-(f^-(t^n, x)) \right) \varphi(t^n, x) \cdot dx \\
&+ \int_{t^n}^{t^{n+1}} \left[ \left( \eta_+(f^+) - \eta_-(f^-) \right) \varphi \right] (t, x_{j+\frac{1}{2}} - 0) \cdot dt \\
&- \int_{t^n}^{t^{n+1}} \left[ \left( \eta_+(f^+) - \eta_-(f^-) \right) \varphi \right] (t, x_{j-\frac{1}{2}} + 0) \cdot dt \leq 0
\end{aligned}$$

for any couple of smooth, convex functions $\eta_\pm \in C^2(\mathbb{R})$ and $j, n \in \mathbb{Z} \times \mathbb{N}$. Clearly, as the Courant number is 1, there is no need for a projection step so the summation on $j, n$ is rather straightforward:

$$- \sum_{j,n \in \mathbb{Z}\times\mathbb{N}} \int_{t^n}^{t^{n+1}} \int_{C_j} \left(\eta_+(f^+) + \eta_-(f^-)\right) \partial_t \varphi + \left(\eta_+(f^+) - \eta_-(f^-)\right) \partial_x \varphi \cdot dx \cdot dt$$

$$\leq \sum_{j \in \mathbb{Z},\, n \in \mathbb{N}} \left( \mathscr{I}^+_{n,j-\frac{1}{2}} - \mathscr{I}^-_{n,j-\frac{1}{2}} \right) \int_{t^n}^{t^{n+1}} \varphi(t, x_{j-\frac{1}{2}}) \cdot dt\,, \tag{5.3}$$

because $\varphi(t,\cdot)$ is continuous in $x = x_{j-\frac{1}{2}}$ and $\varphi(t,\cdot) = 0$ for $t = 0, T$. We used the following notation,

$$\mathscr{I}^\pm_{n,j-\frac{1}{2}} = \eta_\pm \left( f^\pm(t^n, x_{j-\frac{1}{2}} + 0) \right) - \eta_\pm \left( f^\pm(t^n, x_{j-\frac{1}{2}} - 0) \right).$$

These terms $\mathscr{I}^\pm_{n,j-\frac{1}{2}}$ stand for the jump of entropy flux across each zero-wave, located at the grid's interface. They are independent of $t$ thanks to the CFL condition, which ensures that linear waves propagate exactly $\Delta x$ during $\Delta t$.

- One needs to recover, up to $\Delta x$, the source term which appears in the entropy inequality for the exact solution, and which seems to be missing here. By definition of the stationary equations, see (4.16), at any time-step $t^n = n\Delta t$, the corresponding smooth profiles $\tilde{f}^\pm_n$ satisfy modified ODE's too,

$$\partial_x \left( \eta_\pm(\tilde{f}^\pm_n) \right) = k(x) G^\pm(\tilde{f}^+_n, \tilde{f}^-_n), \qquad G^\pm(\tilde{f}^+, \tilde{f}^-) := \eta'_\pm(\tilde{f}^\pm) G(\tilde{f}^+, \tilde{f}^-).$$

Accordingly, the entropy jumps rewrite:

$$\eta_\pm(f^\pm)(t^n, x_{j-\frac{1}{2}} + 0) = \eta_\pm(f^\pm)(t^n, x_{j-\frac{1}{2}} - 0) + \int_{x_{j-1}}^{x_j} k(s) G^\pm\left(\tilde{f}^+_n(s), \tilde{f}^-_n(s)\right) \cdot ds,$$

therefore, the former jumps are amended as follows,

$$\mathscr{I}^+_{n,j-\frac{1}{2}} - \mathscr{I}^-_{n,j-\frac{1}{2}} = \int_{x_{j-1}}^{x_j} k(s) \left( \eta'_+(\tilde{f}^+_n(s)) - \eta'_-(\tilde{f}^-_n(s)) \right) G\left(\tilde{f}^+_n(s), \tilde{f}^-_n(s)\right) \cdot ds.$$

So the contribution of the source term can be reconstructed:

$$\left( \mathscr{I}^+_{n,j-\frac{1}{2}} - \mathscr{I}^-_{n,j-\frac{1}{2}} \right) \int_{t^n}^{t^{n+1}} \varphi(t, x_{j-\frac{1}{2}}) dt$$

$$= \int_{t^n}^{t^{n+1}} \int_{x_{j-1}}^{x_j} k(x) \left[ \eta'_+(f^+) - \eta'_-(f^-) \right] G(f^+, f^-) \varphi(t,x)\, dx \cdot dt \tag{5.4}$$

$$- \int_{t^n}^{t^{n+1}} \int_{x_{j-1}}^{x_j} k(x) \underbrace{\left[ \eta'_+(f^+) - \eta'_-(f^-) \right] G(f^+, f^-)}_{G^+(f^\pm) - G^-(f^\pm) = \beta(t,x)} \left[ \varphi(t,x) - \varphi(t, x_{j-\frac{1}{2}}) \right] dx \cdot dt$$

$$- \int_{t^n}^{t^{n+1}} \varphi(t, x_{j-\frac{1}{2}}) \int_{x_{j-1}}^{x_j} k(x) \Big[ \left(\eta'_+(f^+) - \eta'_-(f^-)\right) G(f^+, f^-)$$

$$\underbrace{- \left(\eta'_+(\tilde{f}_n^+(s)) - \eta'_-(\tilde{f}_n^-(s))\right) G\left(\tilde{f}_n^+(x), \tilde{f}_n^-(x)\right)\Big]}_{G^+(f^\pm)-G^+(\tilde{f}_n^\pm)-G^-(f^\pm)+G^-(\tilde{f}_n^\pm)=\alpha(t,x)} dx \cdot dt.$$

The above terms $\alpha$, $\beta$ are bounded as follows:

$$|\alpha(t,x)| \leq Lip(G)\text{TV}\left(\tilde{f}_n^\pm(\cdot); \{x_{j-1}, x_j\}\right), \quad |\beta(t,x)| \leq |G^+ - G^-| \leq 2C_0.$$

- For any $\ell \in \mathbb{R}$, we approximate a weak Kružkov entropy $u \mapsto |u - \ell|$ by means of a smooth function $E \in C^2(\mathbb{R})$ such that $E'' \geq 0$, $E(v) = |v|$ for $|v| \geq 1$, $E'(0) = 0$ and $|E'| \leq 1$. It is rescaled like $\eta_\delta(v) = \delta E(\frac{v-\ell}{\delta})$, and therefore $\eta'_\delta(v) \to \text{sgn}(v - \ell)$ as $\delta \to 0$, for all $v \neq 0$. Using (5.3) and (5.4) and thanks to the bound above on $\alpha$ and $\beta$, we pass to the limit as $\delta \to 0$ by means of the dominated convergence theorem and finally recover (5.2).

### *5.1.2 Resulting $L^1$ Error Bounds*

Hereafter we shall denote $f^\pm_{\Delta x}$ the piecewise-constant numerical approximations delivered by the WB algorithm formerly described, and keep $f^\pm$ for the corresponding exact solution. Each one satisfies a specific entropy dissipation inequality, (5.1) and (5.2). An error estimate can be derived by taking advantage of the simple fact that (weak) *Kružkov entropies are symmetric*, together with a specific choice of non-negative test-functions. Indeed, adopting the notations of [3, 6],

$$\mathbb{R}^+ \times \mathbb{R} \times \mathbb{R}^+ \times \mathbb{R} \to \mathbb{R}^+, \qquad 0 \leq \phi(t, x, s, y) = \varphi(t, x)\zeta(t - s, x - y).$$

The choice of $\zeta$ corresponds to a smooth approximation of the Dirac mass, namely for $\Delta, \delta > 0$:

$$\zeta(t,x) = \zeta_t(t)\zeta_x(x) = \frac{1}{\delta}\zeta_t^1\left(\frac{t}{\delta}\right) \cdot \frac{1}{\Delta}\zeta_x^1\left(\frac{x}{\Delta}\right), \qquad 0 \leq \zeta_t^1, \zeta_x^1 \in C_0^\infty(\mathbb{R}).$$

Moreover, one can ensure that they are symmetric and:

$$\|\zeta_t\|_{L^1(\mathbb{R})} = \|\zeta_x\|_{L^1(\mathbb{R})} = 1, \qquad \zeta_t^1(\cdot)\zeta_x^1(\cdot) \text{ supported in } (-1, 0) \times \left(-\frac{1}{4}, \frac{1}{4}\right).$$

Now, thanks to entropies' symmetry, it is possible to consider (5.2) with $k_\pm = f^\pm(s, y)$, for any $s, y \in \mathbb{R}^+ \times \mathbb{R}$ and reciprocally. By double integration, and usual simplifications, one arrives at:

$$0 \leq \iiiint \zeta(t-s, x-y)\Big\{\Big[|f^+_{\Delta x}(t,x) - f^+(s,y)| + |f^-_{\Delta x}(t,x) - f^-(s,y)|\Big]\partial_t\varphi(t,x)$$
$$+ \Big[|f^+_{\Delta x}(t,x) - f^+(s,y)| - |f^-_{\Delta x}(t,x) - f^-(s,y)|\Big]\partial_x\varphi(t,x)$$
$$+ \Big(\mathrm{sgn}(f^+_{\Delta x}(t,x) - f^+(s,y)) - \mathrm{sgn}(f^-_{\Delta x}(t,x) - f^-(s,y))\Big)$$
$$\times \Big[k(x)G(f^+_{\Delta x}, f^-_{\Delta x})(t,x) - k(y)G(f^+, f^-)(s,y)\Big]\varphi(t,x)\Big\}\, ds dy dt dx$$
$$+ C_\alpha \sum_{n,j} \mathrm{TV}\left(\tilde{f}^\pm_n(\cdot); \{x_{j-1}, x_j\}\right) \int dy \int ds \int_{t^n}^{t^{n+1}} \int_{x_{j-1}}^{x_j} k(x)\phi(t, x_{j-\frac{1}{2}}, s, y)\, dt dx$$
$$+ 2C_0 \sum_{n,j} \int dy \int ds \int_{t^n}^{t^{n+1}} \int_{x_{j-1}}^{x_j} k(x)\left|\phi(t,x,s,y) - \phi(t, x_{j-\frac{1}{2}}, s, y)\right| dt dx. \tag{5.5}$$

By imposing that $\varphi(t,x)$ is a regularized characteristic function as in [3, 6] with $\nu = 0$, $\delta = \Delta$, $L = 1$ and $\theta = \Delta/4$, space and time derivatives simplify each other in order to produce

$$\int_{x_1-\frac{\Delta}{2}}^{x_2+\frac{\Delta}{2}} |f^\pm_{\Delta x}(T,x) - f^\pm(T,x)| dx$$
$$\leq \int_{x_1-\frac{\Delta}{2}-T}^{x_2+\frac{\Delta}{2}+T} |f^\pm_{\Delta x}(0,x) - f^\pm(0,x)| dx + 4C\mathrm{TV}(f^\pm(0,.\cdot))\Delta + [\ldots].$$

Now, in contrast with the similar computation in [1], one can get rid of the contribution of $G$ in the term (5.5) by taking advantage of its quasi-monotonicity: in fact, since $\pm\frac{\partial G}{\partial f^\pm} \leq 0$ (see (4.18)) and $\mathrm{sgn}(b)a - |a| \leq 0$ for any $a, b \in \mathbb{R}^2$, we have

$$\Big[\mathrm{sgn}(f^+_{\Delta x}(t,x) - f^+(s,y)) - \mathrm{sgn}(f^-_{\Delta x}(t,x) - f^-(s,y))\Big]$$
$$\cdot \Big[G(f^+_{\Delta x}, f^-_{\Delta x})(t,x) - G(f^+, f^-)(s,y)\Big] \leq 0.$$

Since $k(x) \geq 0$, from the integrand in (5.5) we get a negative term, while the remaining term comes from the difference $k(x) - k(y)$ and is smaller than:

$$\int\int\int\int \varphi(t,x)|k(x) - k(y)|\zeta(t-s, x-y)\, ds dy dt dx$$
$$\leq T \int_x \int_y |k(x) - k(y)|\zeta_x(x-y)\, dx dy$$
$$\leq \frac{T}{\Delta} \int_x \int_{-\frac{\Delta}{4}}^{\frac{\Delta}{4}} |k(x) - k(x+\xi)|\, d\xi dx$$
$$\leq \frac{T}{\Delta}\mathrm{TV}(k) \int_{-\frac{\Delta}{4}}^{\frac{\Delta}{4}} |\xi|\, d\xi = \mathrm{TV}(k) \cdot \frac{\Delta}{16} \cdot T \sup|\varphi|.$$

Above, we used that $|\varphi| \leq 1$ by construction. It is necessary to derive suitable bounds for the error terms:

- Following the construction of [3], $|\partial_x \varphi| \leq C/\Delta$ and this affects the term:

$$\int dy \int ds \sum_{n,j} \int_{t^n}^{t^{n+1}} \int_{x_{j-1}}^{x_j} k(x) \left| \phi(t,x,s,y) - \phi(t, x_{j-\frac{1}{2}}, s, y) \right| \, dtdx$$

$$\leq \sum_{n,j} \int_{t^n}^{t^{n+1}} \int_{x_{j-1}}^{x_j} k(x) \left| \varphi(t,x) - \varphi(t, x_{j-\frac{1}{2}}) \right| \, dtdx \, \leq \, CT \frac{\Delta x}{\Delta} \|k\|_{L^1(\mathbb{R})}.$$

- The other term depends on $\mathrm{TV}(\tilde{f}_n^{\pm}; x_{j-1}, x_j)$, which is bounded by $C_0 \Delta x \|k\|_{L^\infty(\mathbb{R})}$, so one gets:

$$\sum_{n,j} \mathrm{TV}\left\{ \tilde{f}_n^{\pm}(\cdot); (x_{j-1}, x_j) \right\} \int_{t^n}^{t^{n+1}} \int_{x_{j-1}}^{x_j} k(x) \varphi(t, x_{j-\frac{1}{2}}) \, dtdx \leq T \cdot C_0 \Delta x \|k\|_{L^\infty} \|k\|_{L^1}.$$

Since $C_\alpha \leq Lip(G) \leq 2$ (within assumptions (4.6), (4.5)) the inequality reduces to:

$$\int_{x_1 - \frac{\Delta}{2}}^{x_2 + \frac{\Delta}{2}} |f_{\Delta x}^{\pm}(T,x) - f^{\pm}(T,x)| \cdot dx \leq \int_{x_1 - \frac{\Delta}{2} - T}^{x_2 + \frac{\Delta}{2} + T} |f_{\Delta x}^{\pm}(0,x) - f^{\pm}(0,x)| \cdot dx$$
$$+ \frac{4CT\Delta x \|k\|_{L^1}}{\Delta} + 2C_0(T\Delta x)\|k\|_{L^1}\|k\|_{L^\infty}$$
$$+ \Delta \left[ 4C\mathrm{TV}(f^{\pm}(0,.\cdot)) + C_0 \mathrm{TV}(k) T/8 \right].$$

The optimal value for $\Delta$ can be computed by standard ways, and one finds:

$$\int_{x_1 - \frac{\Delta}{2}}^{x_2 + \frac{\Delta}{2}} |f_{\Delta x}^{\pm}(T,x) - f^{\pm}(T,x)| \cdot dx \leq \int_{x_1 - \frac{\Delta}{2} - T}^{x_2 + \frac{\Delta}{2} + T} |f_{\Delta x}^{\pm}(0,x) - f^{\pm}(0,x)| \cdot dx$$
$$+ 2T \left\{ 2\sqrt{2\Delta x \, C_0 \|k\|_{L^1} \left[ \frac{4}{T \cdot C_0} \mathrm{TV}(f^{\pm}(0,.\cdot)) + \frac{\mathrm{TV}(k)}{8} \right]} + \Delta x \, C_0 \|k\|_{L^1} \|k\|_{L^\infty} \right\}.$$

The absolute constant $C$ which is used in [3] is fixed here at 2, [5, Theorem 2]. We have established the second estimate, $\mathscr{E}_2$: the proof of Theorem 4.1 is yet complete.

## 5.2 Practical Comparisons Between $\mathscr{E}_1$ and $\mathscr{E}_2$

### 5.2.1 Comparing the Two Estimates on an Example

Let us show the comparison of the two estimates on an elementary example.

1. Assume first that $\mathrm{TV}\{f_0^{\pm}; (x_1, x_2)\} = 0$: the error estimate (4.12) boils down to

$$\int_{x_1+t}^{x_2-t} |f_{\Delta x}^{\pm}(t,x) - f^{\pm}(t,x)|dx \le \min\left\{\Delta x \cdot \mathscr{E}_1;\ 2t\sqrt{\Delta x} \cdot \mathscr{E}_2\right\}.$$

   Accordingly, there is no error at time $t = 0$. For semi-conductor models like (4.7), initial data usually are $J(t = 0, \cdot) \equiv 0$ and $\rho(t = 0, \cdot) = d$, a piecewise-constant doping profile so its error vanishes, too.
2. The initial Maxwellian gap is fixed to $C_0 = \frac{1}{2}$ so the last term of the time-uniform estimate $\mathscr{E}_1$ cancels.
3. On the coefficient $k$, we assume that $\|k\|_{L^\infty(x_1,x_2)} = 1$ and $\|k\|_{L^1(x_1,x_2)} = C/2$, so that $K = 2$; however, we don't restrict its total variation.

Based on all these assumptions, we can estimate the terms $\mathscr{E}_1$, $\mathscr{E}_2$. It is found that

$$K = 2, \qquad K_0 = \frac{15}{7}, \qquad \mathscr{E}_1 = \frac{C}{2}(2 + K_0)$$

and

$$A(t) \equiv \mathrm{TV}\{k; (x_1, x_2)\}, \qquad 2\mathscr{E}_2 = \frac{C}{2}\left(2\sqrt{\frac{1}{C}\mathrm{TV}\{k; (x_1, x_2)\}} + \sqrt{\Delta x}\right).$$

Therefore, the time-dependent estimate dominates the other one as soon as

$$\boxed{t \ge \sqrt{\Delta x}\,\frac{\mathscr{E}_1}{2\mathscr{E}_2} = \frac{2 + K_0}{1 + 2\sqrt{\frac{\mathrm{TV}\{k;(x_1,x_2)\}}{C\Delta x}}}}.$$

The time-uniform estimate $\mathscr{E}_1$ is sharper in case the relaxation term is multiplied by a small, but oscillating (or at least, displaying areas of strong variation) coefficient.

### 5.2.2 An Elementary Numerical Illustration

Proceeding exactly like in [2], it's possible to measure practically the sensitivity of both well-balanced and time-splitting numerical approximations with respect to the total variation of the coefficient $k(x)$. Accordingly it's necessary to reformulate a

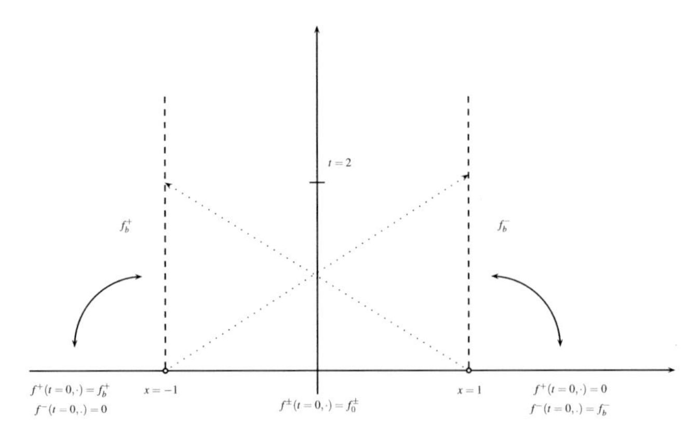


**Fig. 5.1** Conversion of a Cauchy problem in $\mathbb{R}$ into a BVP on $(-1, 1)$

Cauchy problem posed in the whole real line onto initial-boundary value problems (IBVP) on a finite interval, say $x \in (-1, 1)$ in order to achieve computations: see Fig. 5.1. In diagonal variables, (4.8) asks for boundary conditions $f_b^{\pm}$ and initial data $f^{\pm}(t = 0, \cdot) = f_0^{\pm}$. This IBVP is equivalent to the Cauchy problem posed on $x \in \mathbb{R}$ with initial data given by,

$$\begin{cases} f^+(t = 0, x) = f_b^+,\ f^-(t = 0, x) = 0, & \text{for } x \leq -1, \\ f^{\pm}(t = 0, \cdot) = f_0^{\pm}, & \text{for } |x| < 1, \\ f^+(t = 0, x) = 0,\ f^-(t = 0, x) = f_b^-, & \text{for } x \geq 1. \end{cases}$$

Elementary computations based on the scattering matrix written in [4, Lemma 1] furnish the following stationary regimes for $x \in (-1, 1)$:

$$J_\alpha^* = \frac{2\alpha(\bar{e} f_b^+ - f_b^-)}{\bar{e}(1+\alpha) - (1-\alpha)}, \quad \text{where} \quad \bar{e} = \exp(\alpha \|k\|_{L^1(-1,1)}), \tag{5.6}$$

$$f^+(x) = f_b^+ e(x) + \frac{\bar{e} f_b^+ - f_b^-}{\bar{e} - \frac{1-\alpha}{1+\alpha}} (1 - e(x)), \quad \text{where} \quad e(x) = \exp\left(\alpha \int_{-1}^{x} k(y) dy\right), \tag{5.7}$$

$$f^-(x) = f_b^+ e(x) - \frac{\bar{e} f_b^+ - f_b^-}{\bar{e} - \frac{1-\alpha}{1+\alpha}} \left(e(x) - \frac{1-\alpha}{1+\alpha}\right). \tag{5.8}$$

Observe that letting $\alpha \to 0$, one recovers the value $J^*$ found in [2, Sect. 4], i.e. $J_\alpha^* \to J^*$. A sequence of nonnegative, increasingly oscillating, coefficients is now set up,

$$0 \leq k_\mu(x) = \sin^2(\mu \pi x), \qquad \text{for } x \in (-1, 1), \qquad 0 \text{ elsewhere.} \tag{5.9}$$

It has the property that $\|k_\mu\|_{L^1(\mathbb{R})} = 1$, as soon as $\mu \in \mathbb{N}$, so the error estimate $\mathscr{E}_1$, apparently specific for well-balanced schemes, remains invariant with respect to $\mu$ (it depends only of $\|k_\mu\|_{L^1(\mathbb{R})}$). Hence such a behavior should be observed numerically by checking the pointwise errors beyond a certain lapse of time. Oppositely, a time-splitting scheme has its error ruled by an estimate like $\mathscr{E}_2$, obtained by means of entropy dissipation [8], so it should be endowed with a dependency in TV$k_\mu$. In order to actually measure this discrepancy in the simple case where $A(\rho) = \alpha\rho$, we set up the following pointwise errors, for each abscissa $x_j$ in the computational domain, and $t$ large enough:

$$e_\pm^{\Delta t}(t, x_j) = (f^\pm)^{\Delta t}(t, x_j) - f^\pm(x_j), \quad e_\pm^{\Delta x}(t, x_j) = (f^\pm)^{\Delta x}(t, x_j) - f^\pm(x_j), \tag{5.10}$$

where $(f^\pm)^{\Delta x}$ and $(f^\pm)^{\Delta t}$ are piecewise-constant approximations corresponding to a well-balanced and Strang-splitting in time algorithms, respectively. Notice that $(f^\pm)^{\Delta t}$ is therefore formally second-order with respect to time. Numerical results for a simple, formally first-order, time-splitting algorithm are much worse. Hereafter

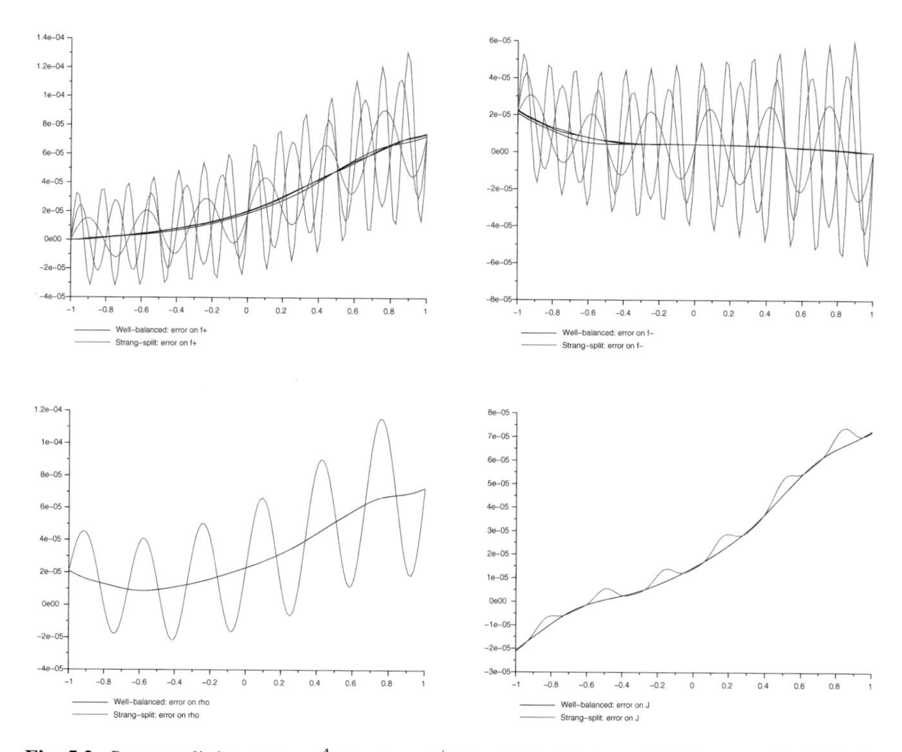


**Fig. 5.2** Strang-splitting errors $e_\pm^{\Delta t}(5,\cdot)$ on $f^+$ (*top, left*), $f^-$ (*top, right*) for $\mu \in \{3, 5, 7\}$; the slowly-varying *black curve* shows the WB error, $e_\pm^{\Delta x}(5,\cdot)$. Corresponding pointwise errors on $\rho(5,\cdot)$ (*bottom, left*) and $J(5,\cdot)$ (*bottom, right*) for $\mu = 3$

we display pointwise errors (5.10) for the values $\alpha = 3, 5, 7$ with,

$$f_b^+ = 1, \quad f_b^- = 0.4, \quad t = 5 \geq 2, \quad \alpha = 0.65, \text{ and } \Delta t = \Delta x = 2^{-6}.$$

An inspection of Fig. 5.2 reveals that:

- As expected, pointwise errors $e_{\pm}^{\Delta x}$ produced by the well-balanced algorithm are insensitive to the growth of $\mathrm{TV}k_\mu$ and remains slowly varying; this confirms the validity of $\mathscr{E}_1$, not depending on $x$-derivatives of the coefficient $k_\mu(x)$.
- Oppositely, oscillations pass from $k_\mu$ into $e_{\pm}^{\Delta}t$ generated by the Strang-split scheme: their amplitude increases with their frequency. There is also a growth of amplitude when going from the left to the right of the computational domain as $\alpha > 0$: if $\alpha < 0$, then oscillations grow in the opposite direction.
- Oscillations pass onto macroscopic quantities $\rho^{\Delta}t$, $J^{\Delta}t$ generated by the Strang-split scheme, too, whereas the ones produced by the WB remain slowly-varying. In particular, the left-right growth still appears on both variables.

## References

1. D. Amadori, L. Gosse, Transient $L^1$ error estimates for well-balanced schemes on non-resonant scalar balance laws. J. Differ. Equ. **255**, 469–502 (2013)
2. D. Amadori, L. Gosse, Error estimates for well-balanced and time-split schemes on a locally damped semilinear wave equation. Math. Comput. http://dx.doi.org/10.1090/mcom/3043
3. F. Bouchut, B. Perthame, Kružkov's estimates for scalar conservation laws revisited. Trans. Am. Math. Soc. **350**(7), 2847–2870 (1998)
4. L. Gosse, A well-balanced scheme able to cope with hydrodynamic limits for linear kinetic models. Appl. Math. Lett. **42**, 15–21 (2015)
5. L. Gosse, Ch. Makridakis, Two a posteriori error estimates for one-dimensional scalar conservation laws. SIAM J. Numer. Anal. **30**, 964–988 (2000)
6. M.A. Katsoulakis, G. Kossioris, Ch. Makridakis, Convergence and error estimates of relaxation schemes for multidimensional conservation laws. Commun. Partial Differ. Equ. **24**(3–4), 395–422 (1999)
7. N.N. Kuznetsov, Accuracy of some approximate methods for computing the weak solutions of a first-order quasilinear equation. Zh. Vychisl. Mat. i Mat. Fiz. **16**, 1489–1502 (1976). English Transl. USSR Comput. Math. Math. Phys. **16**, 105–119 (1976)
8. H.L. Liu, G. Warnecke, Convergence rates for relaxation schemes approximating conservation laws. SIAM J. Numer. Anal. **37**(4), 1316–1337 (2000)

# Chapter 6
# Conclusion and Outlook

**Abstract** In this final chapter we address some problems to which the analysis could be extended. In view of the possible extension of the well-balanced approach to a two-dimensional situation, numerical simulations of two-dimensional Riemann problems for the linear wave equation are shown, together with possible difficulties arising in the application of a Godunov strategy.

**Keywords** Coupling with Poisson equation · 2-dimensional Riemann solver for linear system

So far, many pages were dedicated both to put these "well-balanced" numerical strategies on a firm mathematical ground, and to explain in great detail the reasons why they can outperform on certain classes of problems. This analysis, relying on Bressan-Glimm $L^1$ stability theory for hyperbolic systems of conservation laws, inherits its limitations, which appear to be roughly of two different types:

- Stringent smallness assumptions when genuinely non-linear fluxes are accounted for (left apart the peculiar Temple class).
- Absence of rigorous results for multi-dimensional problems (except for Kružkov's theory of scalar equations and recent ill-posedness results).

Hereafter both these topics will be fairly discussed, and possibly new elements will be displayed which may question the adequacy of Godunov-type numerical strategies when it comes to deal with 2D hyperbolic systems in the presence of vorticity. Besides, observe that kinetic equations, like for instance,

$$\partial_t f(t,x,v) + v\partial_x f = \sigma(x)\left(\int_V f(t,x,v')dv' - f\right),$$

is essentially a shear flow perfectly aligned with the computational grid. If $\sigma \equiv 0$, particles flow at their own velocity $v$, and such a situation can be handled by one-dimensional upwind schemes in a very satisfactory manner. Scattering matrices, as presented in [14, Part II], allow to process the collision term on the right-hand side without generating supplementary errors when $\sigma \neq 0$.



D. Amadori and L. Gosse, *Error Estimates for Well-Balanced Schemes on Simple Balance Laws*, SpringerBriefs in Mathematics,
DOI 10.1007/978-3-319-24785-4_6

## 6.1 Extension to Weakly Nonlinear (Poisson) Coupling

An appealing feature of the time-uniform error estimates stated in Theorem 4.1 lies in the fact that, despite they are a consequence of Glimm-Bressan $L^1$ stability theory, they don't ask for any smallness restriction on the total variation of initial data $f^{\pm}(t=0,\cdot)$. Hence, they hold for "big $BV$ data" with weak relaxation coefficients (small $\|k\|_{L^1}$), making them especially meaningful in the perspective of numerical computations. This robustness with respect to TV $(f^{\pm})$ comes from the semi-linear character of the position-dependent Jin-Xin model (4.1), which allows to prove a time-decay of the Lyapunov functional $\Phi$ (4.57) without constraining interaction potentials in the weights $W_{\pm 1}(x)$. Hopefully, such an advantageous situation may reappear even if the drift coefficient, like the field $E$ in (4.7), derives from a potential which solves Poisson's equation. Clearly, such a weakly nonlinear coupling can bring different levels of difficulty according to its repulsive (Coulomb self-interaction) or attractive (biological confinement, gravitational) character.

### 6.1.1 First Considerations on 1D Poisson Coupling

Within a semi-linear framework, an initial-boundary value problem (IBVP)

$$\begin{cases} \partial_t \rho + \partial_x J = 0, \\ \partial_t J + \partial_x \rho = -\rho \cdot \partial_x \phi - k(x) J, \end{cases} \qquad -\partial_{xx}\phi = \rho, \tag{6.1}$$

where $k(x) \geq 0$, and with constant boundary data for diagonal variables,

$$f^+(t,-1) = f_b^+, \qquad f^-(t,1) = f_b^- \tag{6.2}$$

and the electrostatic potential,

$$\phi(t,-1) = 0, \qquad \phi(t,1) = V \leq 0. \tag{6.3}$$

may be considered as a physically founded hydrodynamic model of semiconductors [19] in case fields remain weak enough. Indeed, three main regimes coexist:

1. A full hydrodynamic one, yielding a quasi-linear, possibly multi-dimensional hyperbolic system coupled to Poisson equation, investigated for instance in [30] (or [22] in a linearized setting).
2. A regime in which inertial effects are neglected over the momentum relaxation time scale. Such an approximation stems on the assumption that carriers' drift energy is much smaller than thermal energy (or that their drift energy is a negligible part of kinetic energy) and is referred to as being "inertial".

3. A regime, contained within the aforementioned, in which temperature variation is considered negligible too, so that heat flux is insignificant. It is characterized as being friction-dominated and mostly referred to as drift–diffusion, [20, 26].

The Poisson equation can be solved in terms of $\rho$, by means of a Green function,

$$\phi(t,x) = \frac{V}{2}(x+1) + \frac{1}{2}\int_{-1}^{1} G(x,y)\rho(y,t)\,dy \tag{6.4}$$

with

$$G(x,y) = \begin{cases} (y+1)(1-x) & \text{if} \quad -1 < y < x \\ (x+1)(1-y) & \text{if} \quad x < y < 1, \end{cases}$$

so that

$$\begin{aligned} \partial_x\phi(t,x) &= \frac{V}{2} + \frac{1}{2}\int_{-1}^{1} \partial_x G(x,y)\rho(t,y)\,dy \\ &= \frac{V}{2} - \frac{1}{2}\int_{-1}^{x} (y+1)\rho(t,y)\,dy - \frac{1}{2}\int_{x}^{1} (y-1)\rho(t,y)\,dy. \end{aligned} \tag{6.5}$$

In diagonal form, (6.1) rewrites as

$$\begin{cases} \partial_t f^- - \partial_x f^- = \frac{1}{2}\left((\partial_x\phi + k)f^+ + (\partial_x\phi - k)f^-\right), \\ \partial_t f^+ + \partial_x f^+ = -\frac{1}{2}\left((\partial_x\phi + k)f^+ + (\partial_x\phi - k)f^-\right), \end{cases} \qquad -\partial_{xx}\phi = \rho. \tag{6.6}$$

This being said, since the convenient sub-characteristic condition reads,

$$\forall t, x \in \mathbb{R}^+ \times (-1,1), \qquad |\partial_x\phi(t,x)| < k(x),$$

it is more natural to let non-negative coefficients appear in the right-hand side:

$$\begin{cases} \partial_t f^\pm \pm \partial_x f^\pm = \pm\frac{1}{2}\left((k - \partial_x\phi)f^- - (k + \partial_x\phi)f^+\right). \\ \qquad -\partial_{xx}\phi = f^+ + f^- = \rho. \end{cases} \tag{6.7}$$

Hopefully, (6.7) may be the most straightforward extension of position-dependent sub-characteristic relaxation models, as considered in Chap. 4, still allowing to produce original $L^1$ error bounds, for WB schemes, through Lyapunov techniques.

### 6.1.2 Derivation of a Stationary Exact Solution

Simplifying assumptions, $\phi(x=-1) = \phi(x=1)$, $\phi(x=0) = 0$ and $\phi$ is an even function of $x$ meet with the standard situation with a semi-conductor with no doping profile and zero volt at its edges. So, $\rho$ is an even function of $x$, too, and

$J = 0$, meaning that there's no current flowing inside the device. Accordingly, steady macroscopic variables satisfy in $x \in (-1, 1)$,

$$\partial_x J = 0 \text{ and } J = 0, \qquad \partial_x \rho = -\rho \partial_x \phi, \qquad -\partial_{xx}\phi = \rho.$$

Since $\frac{\partial_x \rho}{\rho} = -\partial_x \phi$, it comes that $\rho(x) = \exp(\phi(-1) - \phi(x))\rho(-1)$. Let's call

$$\rho_0 := \exp(\phi(-1))\rho(-1) = \exp(\phi(-1))(f_b^+ + f_b^-),$$

where $f_b^{\pm}$ stand for incoming values at $x = \mp 1$, respectively. As $J = 0$,

$$\rho(x = -1) = 2f_b^+ = \rho(x = 1) = 2f_b^- = f_b^+ + f_b^-.$$

The ODE on $\phi$ is a steady Liouville wave equation,

$$\frac{d^2\phi}{dx^2} = \exp(\phi)\rho_0, \qquad \phi \text{ even in } x.$$

By multiplying on both sides by $\partial_x \phi$, one derives,

$$|\partial_x \phi|^2 = 2\partial_x \phi \exp(\phi)\rho_0 = 2\rho_0 \partial_x[\exp(\phi)].$$

Since $\phi(x)$ is even, signs must be carefully monitored when removing the modulus,

$$\pm \frac{\partial_x \phi}{\sqrt{2\rho_0(\exp(\phi) - 1)}} = 1 \qquad \text{because } \phi(x = 0) = 0.$$

In the repulsive case, $\partial_x \phi$ has the same sign as $x$ and there's no integration constant because $\phi$ is even

$$\int \frac{d\phi}{\sqrt{\exp(\phi) - 1}} = x\sqrt{2\rho_0} = 2 \arctan\left(\sqrt{\exp(\phi) - 1}\right),$$

on the positive branch. Then, by easy computations,

$$\sqrt{\exp(\phi) - 1} = \tan^2\left(x\sqrt{\rho_0/2}\right), \qquad \phi(x) = \log\left[1 + \tan^2\left(x\sqrt{\rho_0/2}\right)\right].$$

For instance, with $\rho_0 = 2$, one computes easily that, (see Fig. 6.1)

$$\frac{d^2}{dx^2}\log(1 + \tan^2 x) = 2\sec^2 x = \frac{2}{\cos^2 x} = 1 + \tan^2 x = \exp(\log(1 + \tan^2 x)),$$

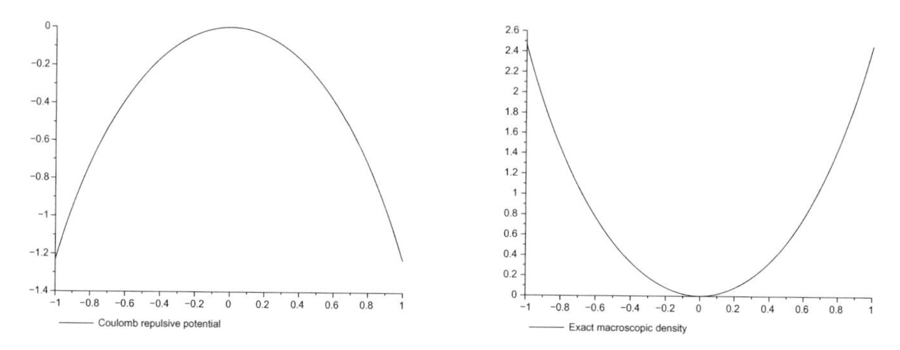


**Fig. 6.1** Exact "Coulomb" stationary solutions $\phi$ (*left*), $\rho$ (*right*) with $\rho_0 = 2$

and the ODE is well satisfied. The real Coulomb repulsive potential is

$$\boxed{\phi(x) = -\log\left[1 + \tan^2\left(x\sqrt{\rho_0/2}\right)\right], \qquad \rho(x) = \rho_0 \cdot \left[1 + \tan^2\left(x\sqrt{\rho_0/2}\right)\right].}$$

To treat attractive potentials too, it suffices to change $\tan^2(\cdot)$ into $\tanh^2(\cdot)$.

## 6.2 A Glimpse on the Difficulty of 2D Interactions

According to [3], defining a reliable notion of well-balanced in several space dimensions isn't obvious. A recent way out of this bad situation was proposed in [6], by means of an astute, purely algebraic, reformulation of existing one-dimensional schemes in terms of the adjoint equation. Such a new setup extends to bidimensional problems in certain cases, and it appears sufficient to have at hand a reliable Godunov-type scheme for the underlying homogeneous system in order to activate this program. A paradigm for 2D hyperbolic systems is the linear wave equation,

$$\boxed{\partial_t p + \nabla \cdot \mathbf{u} = 0, \qquad \partial_t \mathbf{u} + \nabla p = 0,} \tag{6.8}$$

where $p(t, x, y)$ stands for a scalar fluctuation of pressure, and the vector $\mathbb{R}^2 \ni \mathbf{u} = (u, v)^T$ indicates the local velocity of acoustic waves.

### *6.2.1 Exact Kirchhoff Solution of the 2D Riemann Problem*

In the realm of Godunov-type approximations, the Riemann problem is of fundamental importance, because it is the main building block in case one aims at deriving a numerical scheme of "transport-projection" type. Yet, it happens that even for a linear system like (6.8), the structure of a Riemann solution, still being self-similar, is much more intricate compared to the one-dimensional one. To get a flavor of it, let's assume the solution is smooth enough to differentiate $p$ in $t$, and $\mathbf{u}$ in space,

$$\partial_{tt}p - \Delta p = 0, \qquad \Delta \text{ the Laplace operator in } x, y,$$

with identical initial data $p(t=0,\cdot,\cdot)$, but $\partial_t p(t=0,\cdot,\cdot) = -\nabla\cdot\mathbf{u}(t=0,\cdot,\cdot)$, which in the Riemann case, is generally a Dirac measure. A fundamental property of (6.8) is "vorticity preservation", in the sense that,

$$\partial_t\omega(t,x,y) = 0, \qquad \omega = \partial_y u - \partial_x v.$$

These properties imply that, thanks to Kirchhoff's formulas in 2D, one can retrieve the exact expression of the solution of a 2D Riemann problem: this was achieved in the papers [10–12, 24]. By linearity, it is sufficient to restrict initial data to be zero everywhere except in the quadrant $x>0$, $y>0$; in this case, denoting $p_0, u_0, v_0$ the corresponding initial values, and the self-similarity variables $\xi = \frac{x}{t}$, $\eta = \frac{y}{t}$,

$$\begin{cases} p(\xi,\eta) = \frac{p_0}{2\pi}\arccos\frac{\xi\eta}{\sqrt{(1-\xi^2)(1-\eta^2)}} - \frac{u_0}{2\pi}\arccos\frac{-\eta}{\sqrt{1-\xi^2}} - \frac{v_0}{2\pi}\arccos\frac{-\xi}{\sqrt{1-\eta^2}}, \\ u(\xi,\eta) = \frac{p_0}{2\pi}\arccos\frac{-\eta}{\sqrt{1-\xi^2}} + \frac{u_0}{2\pi}\arctan\frac{\xi\sqrt{1-\xi^2-\eta^2}}{\eta} + \frac{v_0}{2\pi}\operatorname{argch}\frac{1}{\sqrt{\xi^2+\eta^2}}, \\ v(\xi,\eta) = \frac{p_0}{2\pi}\arccos\frac{-\xi}{\sqrt{1-\eta^2}} + \frac{u_0}{2\pi}\operatorname{argch}\frac{1}{\sqrt{\xi^2+\eta^2}} + \frac{v_0}{2\pi}\arctan\frac{\eta\sqrt{1-\xi^2-\eta^2}}{\xi}, \end{cases} \tag{6.9}$$

inside the Mach cone, $\xi^2+\eta^2 \le 1$. Since the steady-state equation for $p$ is elliptic, there's no particular direction of propagation emerging in (6.9): thus, postulating numerical fluxes along the computational grid's directions is much more questionable than in one space dimension where Riemann problems always display self-similar elementary waves with neat directions. Moreover, the "radial velocity terms" in $(\xi^2+\eta^2)^{-\frac{1}{2}}$ blow up in $x=y=0$, so the Riemann pattern is unbounded as soon as $u_0 \neq 0$ or $v_0 \neq 0$. Oppositely, the solution $p(\xi,\eta)$ is always continuous.
As a first illustration of (6.9), the solution of (6.8) with curl-free data,

$$p(t=0,x,y) = \chi_{\xi>0}\chi_{\eta>0}, \qquad p_0 = 1, \qquad u_0 = v_0 = 0.$$

is displayed on Fig. 6.2: taking local averages of such a smooth and well-partitioned pattern in order to derive numerical flux functions seems perfectly reasonable.
Building on the system's linearity, a slightly more delicate problem can be solved,

$$p(t=0,x,y) = \operatorname{sgn}(xy), \qquad p_0 = 1, \qquad u_0 = v_0 = 0.$$

Figure 6.3 displays compressive dynamics where high pressure areas endowed with initially null velocity evolve into both a depletion and a fast diagonal flow.

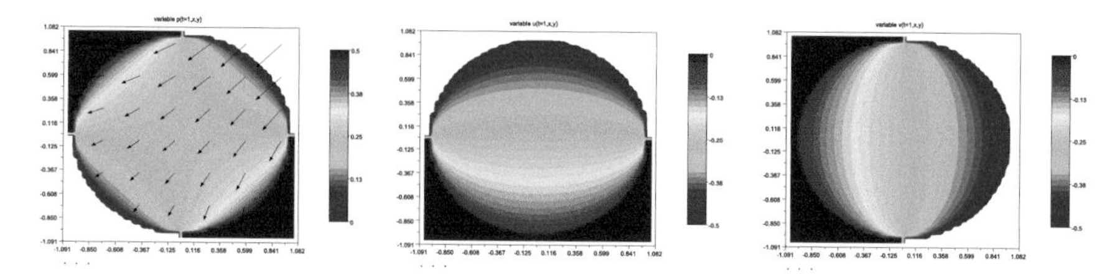

**Fig. 6.2** Riemann pattern (6.9) at $t = 1$ with $p_0 = 1$ and $u_0 = v_0 = 0$

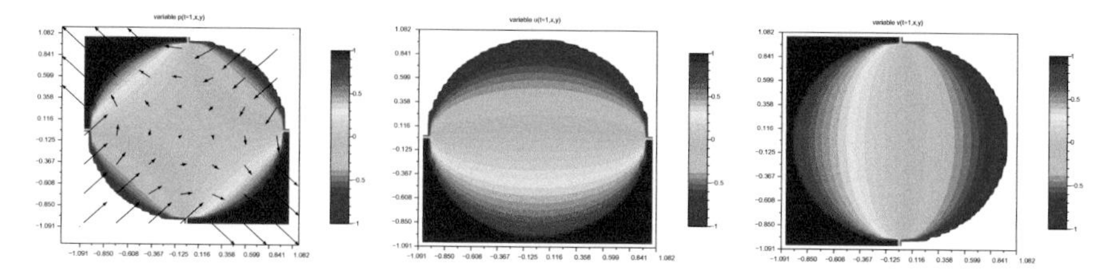

**Fig. 6.3** Riemann pattern at $t = 1$ with $p(t = 0, x, y) = \text{sgn}(xy)$ and $u_0 = v_0 = 0$ (null vorticity)

### 6.2.2 Derivation of the Expression of $p(\xi, \eta)$

We seek an explicit solution of (6.8) with special Riemann-type initial data:

$$(p, u, v)(0, x, y) = \begin{cases} (p_0, u_0, v_0) & x > 0,\ y > 0 \\ (0, 0, 0) & \text{otherwise} \end{cases} \tag{6.10}$$

for generic constants $p_0, u_0, v_0$. Clearly, $p$ satisfies the linear wave equation in

$$D = \{(x, y) : \ x > 0, y > 0\},$$

with discontinuous/measure data,

$$\begin{aligned} p(0, x, y) &= p_0 \cdot \chi_D(x, y)\,, \\ \partial_t p(t = 0, \cdot, \cdot) &= -\nabla \cdot \mathbf{u}(t = 0, \cdot, \cdot) = -u_0 \delta_{x=0, y>0} - v_0 \delta_{x>0, y=0}. \end{aligned} \tag{6.11}$$

**Proposition 6.1** *Inside the open disc $x^2 + y^2 < t^2$, the Riemann solution $p(t, x, y)$ generated from (6.11) is self-similar and reads like (6.9) for $\xi = x/t$ and $\eta = y/t$,*

*Proof* For generic initial data (here expressed in polar coordinates $\rho, \theta$),

$$\tilde{p}(0, \rho, \theta) = p_0(\theta)\,, \qquad \partial_t \tilde{p}(0, \rho, \theta) = \frac{1}{\rho} p_1(\theta).$$

we have Kirchhoff-Poisson formula:

$$p(t,x,y) = \frac{1}{2\pi} I_0(t,x,y) + \frac{1}{2\pi} I_1(t,x,y),$$

with

$$I_0(t,x,y) = \partial_t \left( \iint_{B_t(x,y)} \frac{p_0(x',y')}{\sqrt{t^2-(x'-x)^2-(y'-y)^2}} \, dx'dy' \right)$$
$$I_1(t,x,y) = \iint_{B_t(x,y)} \frac{p_1(x',y')}{\sqrt{t^2-(x'-x)^2-(y'-y)^2}} \, dx'dy'$$

where $B_t(x,y)$ stands for the ball of radius $t$ centered at $(x,y)$. Since $x^2+y^2=r^2<t^2$, each domain of integration in $I_0$, $I_1$ contains the origin. In polar coordinates: $(x,y)=r\mathrm{e}^{i\theta}$ and $(x',y')=\rho\mathrm{e}^{i\alpha}$, with $\alpha \in [0,2\pi)$ and $0<\rho\le\rho^+(\alpha)$ where

$$\|\rho^+(\alpha)\mathrm{e}^{i\alpha} - r\mathrm{e}^{i\theta}\| = t \quad \Rightarrow \quad \rho^+(\alpha) = r\cos(\alpha-\theta) + \sqrt{t^2 - r^2\sin^2(\alpha-\theta)}.$$

Moreover we get

$$(x'-x)^2+(y'-y)^2 = \|\rho\mathrm{e}^{i\alpha} - r\mathrm{e}^{i\theta}\|^2 = \|\rho\mathrm{e}^{i(\alpha-\theta)} - r\|^2 = \rho^2 + r^2 - 2r\rho\cos(\alpha-\theta),$$

hence

$$I_0(t,\rho,\theta) = \partial_t \left[ \int_0^{2\pi} d\alpha \, p_0(\alpha) \left( \int_0^{\rho^+(\alpha)} \frac{\rho}{\sqrt{t^2-r^2-\rho^2+2r\rho\cos(\alpha-\theta)}} \, d\rho \right) \right] \tag{6.12}$$

$$I_1(t,\rho,\theta) = \int_0^{2\pi} d\alpha \, p_1(\alpha) \left( \int_0^{\rho^+(\alpha)} \frac{1}{\sqrt{t^2-r^2-\rho^2+2r\rho\cos(\alpha-\theta)}} \, d\rho \right). \tag{6.13}$$

These integrals in $d\rho$ above are improper, since the integrand is unbounded of order $\frac{1}{2}$ as $\rho \to \rho^+$. For later use, we extract the following expression from (6.12) and (6.13):

$$\begin{aligned} t^2-r^2-\rho^2+2r\rho\cos(\alpha-\theta) &= \left[t^2 - r^2\sin^2(\alpha-\theta)\right] - [\rho - r\cos(\alpha-\theta)]^2 \\ &= A^2 - [\rho - B]^2, \end{aligned}$$

with

$$A = \sqrt{t^2 - r^2\sin^2(\alpha-\theta)}, \qquad B = r\cos(\alpha-\theta). \tag{6.14}$$

Notice that $\rho^+(\alpha) = A + B$ and that $A^2 - B^2 = t^2 - r^2$.

- Computation of integral (6.13):

$$\int_0^{\rho^+} \frac{1}{\sqrt{t^2 - r^2 - \rho^2 + 2r\rho\cos(\alpha - \theta)}}\, d\rho = \int_0^{B+A} \frac{1}{\sqrt{A^2 - [\rho - B]^2}}\, d\rho$$
$$= -\left[\arccos\left(\frac{\rho - B}{A}\right)\right]_0^{\rho^+} = -\arccos\left(\frac{\rho^+ - B}{A}\right) + \arccos\left(\frac{-B}{A}\right)$$
$$= -\arccos(1) + \arccos\left(\frac{-B}{A}\right) = \arccos\left(\frac{-B}{A}\right)$$
$$= \arccos\left(\frac{-r\cos(\alpha - \theta)}{\sqrt{t^2 - r^2\sin^2(\alpha - \theta)}}\right),$$

so that we obtain for (6.13):

$$I_1(t, \rho, \theta) = \int_0^{2\pi} p_1(\alpha) \arccos\left(\frac{-r\cos(\alpha - \theta)}{\sqrt{t^2 - r^2\sin^2(\alpha - \theta)}}\right) d\alpha.$$

In particular, if $p_1 = -u_0\delta_{\alpha=\pi/2} - v_0\delta_{\alpha=0}$, it comes

$$I_1(t, \rho, \theta) = -u_0 \arccos\left(\frac{-r\cos(\frac{\pi}{2} - \theta)}{\sqrt{t^2 - r^2\sin^2(\frac{\pi}{2} - \theta)}}\right) - v_0 \arccos\left(\frac{-r\cos(-\theta)}{\sqrt{t^2 - r^2\sin^2(-\theta)}}\right)$$
$$= -u_0 \arccos\left(\frac{-r\sin\theta}{\sqrt{t^2 - r^2\cos^2\theta}}\right) - v_0 \arccos\left(\frac{-r\cos\theta}{\sqrt{t^2 - r^2\sin^2\theta}}\right)$$
$$= -u_0 \arccos\left(\frac{-\eta}{\sqrt{1 - \xi^2}}\right) - v_0 \arccos\left(\frac{-\xi}{\sqrt{1 - \eta^2}}\right)$$

being $\xi = x/t = (r/t)\cos\theta$ and $\eta = y/t = (r/t)\sin\theta$.

- Computation of integral (6.12):

$$\int \frac{\rho}{\sqrt{A^2 - [\rho - B]^2}}\, d\rho = \int \frac{\rho - B}{\sqrt{A^2 - [\rho - B]^2}}\, d\rho + B\int \frac{1}{\sqrt{A^2 - [\rho - B]^2}}\, d\rho$$
$$= -\sqrt{A^2 - [\rho - B]^2} - B\arccos\left(\frac{\rho - B}{A}\right),$$

so that

$$
\begin{aligned}
\int_0^{\rho^+} & \frac{\rho}{\left(t^2 - r^2 - \rho^2 + 2r\rho\cos(\alpha-\theta)\right)^{1/2}}\,d\rho = \int_0^{B+A} \frac{\rho}{\left(A^2 - [\rho - B]^2\right)^{1/2}}\,d\rho \\
&= -\left[\sqrt{A^2 - [\rho - B]^2}\right]_0^{A+B} - B\left[\arccos\left(\frac{\rho - B}{A}\right)\right]_0^{A+B} \\
&= \sqrt{A^2 - B^2} + B\arccos\left(\frac{-B}{A}\right).
\end{aligned}
$$

Recalling the definitions (6.14) of $A$ and $B$,

$$
\begin{aligned}
\partial_t & \left(\int_0^{\rho^+} \frac{\rho}{\left(t^2 - r^2 - \rho^2 + 2r\rho\cos(\alpha-\theta)\right)^{1/2}}\,d\rho\right) \\
&= \partial_t\left(\sqrt{A^2 - B^2}\right) + B\partial_t\left(\arccos\left(\frac{-B}{A}\right)\right) \\
&= \frac{A^2}{\sqrt{A^2 - B^2}}\frac{\partial_t A}{A} - \frac{B^2}{\sqrt{A^2 - B^2}}\frac{\partial_t A}{A} = \frac{\partial_t A}{A}\sqrt{A^2 - B^2} \\
&= \frac{t}{t^2 - r^2\sin^2(\alpha-\theta)}\sqrt{t^2 - r^2}.
\end{aligned}
$$

Going back to (6.12),

$$
\begin{aligned}
I_0(t,\rho,\theta) &= t\sqrt{t^2 - r^2}\int_0^{2\pi} p_0(\alpha)\frac{1}{t^2 - r^2\sin^2(\alpha-\theta)}d\alpha \\
&= t\sqrt{t^2 - r^2}\, p_0 \int_0^{\pi/2} \frac{1}{t^2 - r^2\sin^2(\alpha-\theta)}d\alpha \\
&= \sqrt{1 - \frac{r^2}{t^2}}\, p_0 \int_0^{\pi/2} \frac{1}{1 - \frac{r^2}{t^2}\sin^2(\alpha-\theta)}d\alpha \\
&= \sqrt{1 - \frac{r^2}{t^2}}\, p_0 \frac{t^2}{r^2}\int_0^{\pi/2} \frac{1}{\frac{t^2}{r^2} - \sin^2(\alpha-\theta)}d\alpha.
\end{aligned}
$$

Since for $C = t^2/r^2 > 1$

$$
\int \frac{1}{C - \sin^2 x}\,dx = \frac{1}{\sqrt{C(C-1)}}\arctan\left(\sqrt{\frac{C-1}{C}}\tan x\right),
$$

then

$$
\begin{aligned}
I_0(t,\rho,\theta) &= p_0\sqrt{1-\frac{1}{C}}\,\frac{C}{\sqrt{C(C-1)}}\left[\arctan\left(\sqrt{1-\frac{r^2}{t^2}}\tan(\alpha-\theta)\right)\right]_{\alpha=0}^{\alpha=\pi/2}\\
&= p_0\left[\arctan\left(\sqrt{1-\frac{r^2}{t^2}}\tan(\pi/2-\theta)\right)-\arctan\left(\sqrt{1-\frac{r^2}{t^2}}\tan(-\theta)\right)\right]\\
&= p_0\left[\arctan\left(\sqrt{1-\frac{r^2}{t^2}}\frac{1}{\tan\theta}\right)+\arctan\left(\sqrt{1-\frac{r^2}{t^2}}\tan\theta\right)\right]\\
&= p_0\arctan\left(\frac{t^2}{r^2}\sqrt{1-\frac{r^2}{t^2}}(\cot\theta+\tan\theta)\right)=p_0\arctan\left(\frac{t^2}{xy}\sqrt{1-\frac{r^2}{t^2}}\right)
\end{aligned}
$$

thanks to the “arctangent addition formula”. Further, for any $z\in(-\pi,\pi)$,

$$
\arctan x+\arctan y=\arctan\frac{x+y}{1-xy},\qquad \cos(\arctan z)=\frac{1}{\sqrt{1+z^2}},
$$

so that this last expression expression reduces to,

$$
\frac{I_0}{p_0}=\arccos\sqrt{\frac{x^2y^2}{t^4-t^2(x^2+y^2)+x^2y^2}}=\arccos\frac{\xi\eta}{\sqrt{(1-\xi^2)(1-\eta^2)}}.
$$

Yet, $u,v$ follow by differentiating $p$ in $x$ or $y$, respectively, then integrating in time according to Rankine-Hugoniot conditions on the circle $x^2+y^2=t^2$, [5, 11, 24].

### 6.2.3 Dirac Vorticity and the Mach Cone Structure

The structure displayed on Fig. 6.2 is specific to data for which $\omega(t,\cdot,\cdot)\equiv 0$. In order to see what happens when the vorticity doesn't vanish, one can follow benchmarks partly suggested in [21, 23, 29] and study the family of problems given by:

$$
p(t=0,x,y)=0,\quad u(t=0,x,y)=\begin{cases}0\\ \frac{\operatorname{sgn}(xy)}{2}\\ \operatorname{sgn}(xy)\\ \operatorname{sgn}(xy)\\ \operatorname{sgn}(xy)\end{cases},\quad v(t=0,x,y)=\begin{cases}\operatorname{sgn}(xy)\\ \operatorname{sgn}(xy)\\ \operatorname{sgn}(xy)\\ \frac{\operatorname{sgn}(xy)}{2}\\ 0.\end{cases}
\tag{6.15}
$$

As the data is supported on each of the four quadrants, it is yet necessary to exploit linearity in order to split it into more elementary Riemann problems supported on one quadrant only, then to rotate of $\pm\frac{\pi}{2}$ or $\pi$ and change signs of $u$ or $v$ in order to be able to use (6.9), see Fig. 6.4. Exact solutions obtained in this manner are

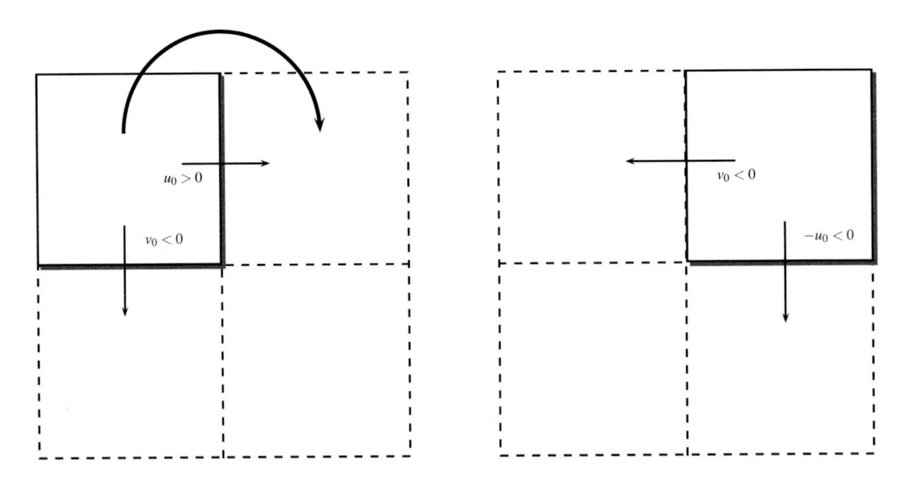


**Fig. 6.4** Switching of velocities in order to solve general Riemann data with (6.9)

displayed on Fig. 6.5. The situation is now completely different as the structure of the velocity vector $u, v$ is much more involved. On the first line, the radial structure blowing up at the origin appears neatly, this corresponds to the figures shown in [21]. The third line refers to the benchmarks presented in [23, 29]: the velocity vector appears both to swirl around the origin and to keep track of some (not all) of the initial discontinuities. The third line shows an even more intricate picture: only the "external waves", located on the edges $x = \pm 1, y = \pm 1$ appear to be usable in order to practically compute numerical fluxes. Another salient feature of Fig. 6.5 is to reveal the smooth deformations of the structure in the Mach cone when initial intensities of horizontal and vertical velocities switch to one another. Corresponding vorticities are preserved: in the first and last case of (6.15),

$$\begin{cases} \omega(t,x,y) = -\partial_x v(t=0,x,y) = -2\delta(x)\text{sgn}(y), \\ \omega(t,x,y) = \partial_y u(t=0,x,y) = 2\text{sgn}(x)\delta(y), \end{cases}$$

whereas in the third one, for which $u(t=0,x,y) = v(t=0,x,y)$,

$$\omega(t,x,y) = \partial_y u(t=0,x,y) - \partial_x v(t=0,x,y) = 2\delta(y)\text{sgn}(x) - 2\delta(x)\text{sgn}(y).$$

Accordingly, Fig. 6.6 reveals that the Riemann patterns derived from (6.9) are such that these values of $\omega$ are kept exactly. Intricate swirling velocities appear to be related to such singular (measure-valued) vorticities.

An inspection of 2D Riemann patterns, like the ones displayed on Fig. 6.5, reveals that the derivation of "multidimensional flux functions" within a Godunov scheme by means of Stokes theorem will be a quite difficult task.

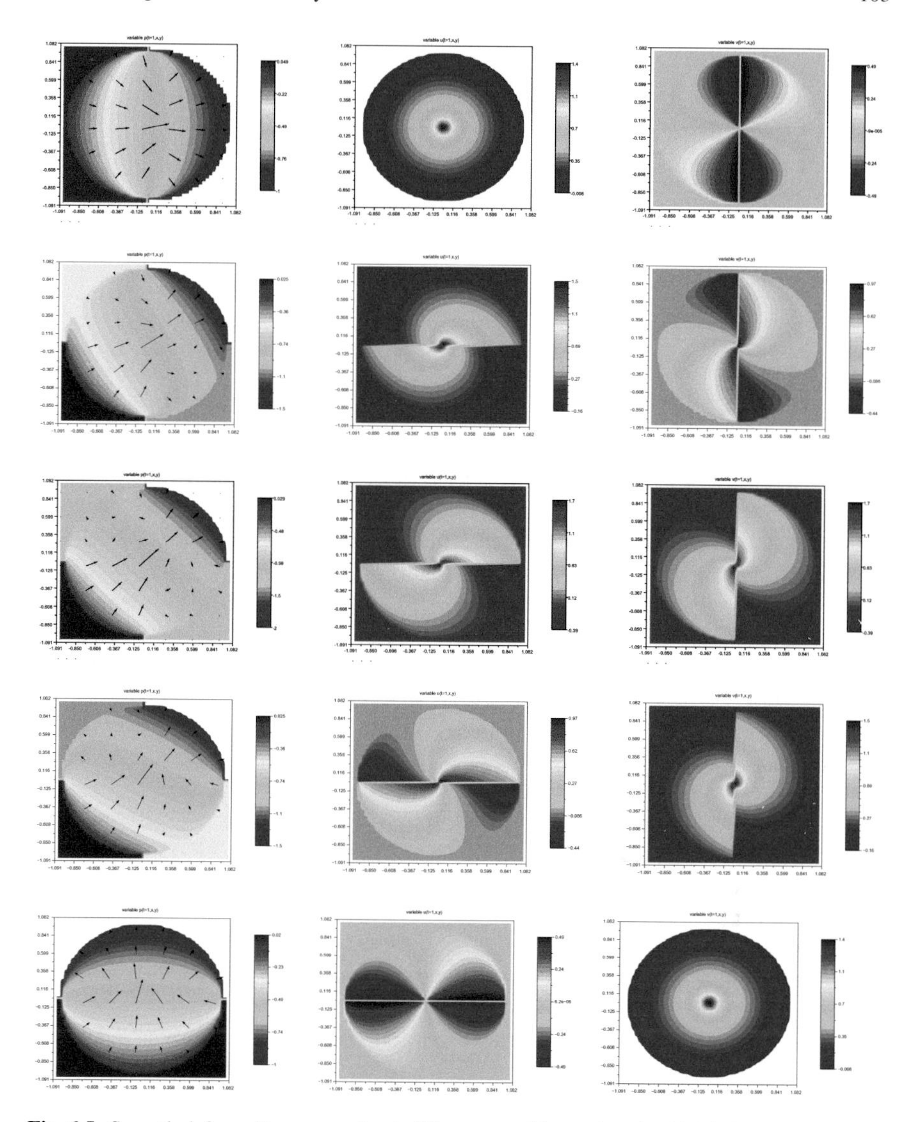

**Fig. 6.5** Smooth deformations occurring in Riemann problems given by (6.15) for (6.8)

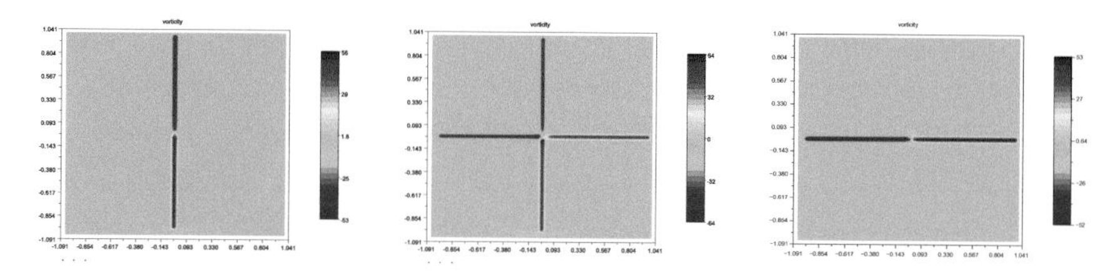

**Fig. 6.6** Vorticity: $u(t=0)=0$ (*left*), $u(t=0)=v(t=0)=\mathrm{sgn}(xy)$ (*middle*), $v(t=0)=0$ (*right*)

Indeed, like in the 2D scalar case [15], flux functions result from a double integration, in both space and time, of varying quantities; oppositely, in 1D, one just has to integrate in time a constant value, a trivial computation. Accordingly, in a 2D context, the Godunov scheme may be better apprehended by coming back to its original formulation, which consists in locally averaging Riemann patterns, the viewpoint advocated in [33] with an HLLE solver.

### 6.2.4 Artificial Vorticity and Shortcomings of Riemann Solvers

The essence of Godunov's time-marching scheme is to proceed by iterations of "transport-projection": initial data are first projected onto a subspace of functions constant on each computational cell, this operation defines a sequence of 2D Riemann problems which are meant to be solved for a short time-step $\Delta t > 0$, after which corresponding patterns are again projected on cell-wise constant functions, a restarting procedure. Obviously, in the vast majority of cases, Dirac-type vorticity pops up or remains after each projection step, which results in the previously seen swirling Riemann patterns; more seriously, this happens even if the initial data was chosen curl-free, so that the solution's vorticity must remain zero for all times. The situation is quite different compared to what happens in one space dimension,

- On the one hand, initial data for 2D Riemann problems may be discontinuous, so the Riemann problem is a necessary building block in order to evolve correctly in time these jumps, at the price of possibly having to take local averages of fine-scale structures. Oppositely to 1D, piecewise constant datasets in 2D are "approximately stable", under the action of (6.8), only if vorticity is null.
- But on the other hand, the projection operator induces inevitably Dirac-type vorticity, supported on the "skeleton" of the computational grid. That potentially spurious vorticity cannot decay in the evolution step as (6.8) preserves it. Worse, it is thought that, for quasi-linear systems, one should expect neither uniqueness, nor stability in the presence of vorticity, [9] (see also [2, 8, 25]).
- When working on a Cartesian uniform grid, all the Riemann problems are set at the cells' vertices. It may happen that a shear flow appears along a line which isn't parallel to one of the axes, so it is projected as a "staircase" on the grid, see Fig. 6.7. This ignites a spurious mechanism which spreads the shear line over many cells, as was shown in [31, 32] for simple dimensional-splitting schemes.

So, in some sense, the projection step appears to be a cause of dysfunction of the Riemann solver because of the singular vorticity it creates even if initial data are very smooth or curl-free. There is some inadequacy between both the processes involved in the Godunov scheme, even in such a favorable context where an exact Riemann solver appears to be available. Perhaps Lax-Friedrichs solvers on staggered

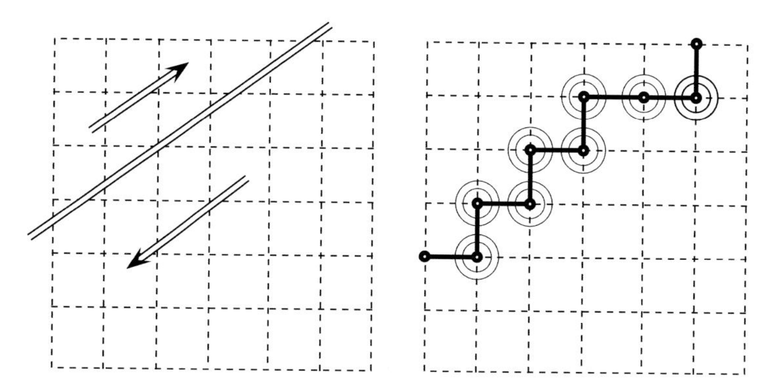

**Fig. 6.7** Breakup of an oblique shear flow and resulting mach cones ("vortex patches")

computational grids [4] may work better as they rely only on "external waves", that is to say, the one-dimensional wave fans which occur outside of the Mach cone, away from those intricate interactions. Anyway, it is still a delicate question to devise a well-balanced scheme in 2D, because no really satisfying numerical algorithm seems to be at hand, even for the treatment of homogeneous systems like (6.8).

## 6.3 Putting Our Results in Perspective

Several open problems and extensions were already evoked. Others may include:

- Deriving new error estimates by means of Lyapunov functional for 1D systems of balance laws endowed with stronger nonlinearities, but keeping necessary properties like invariant regions or $BV$ bounds. This encompasses (for instance) systems of two equations belonging to the Temple's class; an interesting example is motivated by traffic flow modeling, see [1, Sect. 3.3].
- Proving $L^2$ error estimates for relativistic wave equations on curved spacetimes, see e.g. [16, 17], exploiting a decreasing "Bony functional" presented in [35].
- Implementing and testing a "genuinely 2D Godunov scheme" based on Kirchhoff-based Riemann solvers, probably following [33], that is, without resorting to flux functions formalism. This may be useful in order to cope with the new multi-dimensional well-balanced ideas published in [6], especially for applications to diffusive relaxation of damped wave equations.
- Hyperbolic relaxation schemes for 2D conservation laws, already evoked in [18], but now relying on an actual 2D wave equation instead of two one-dimensional processes within a dimensional-splitting procedure. In its "relaxed version", it would simply consist in collapsing $u_0 = f(p_0)$, $v_0 = g(p_0)$ at each time-step in (6.9). This may bring improved accuracy in relaxation approximations of 2D scalar conservations endowed with intricate flux functions $f, g$ for which multi-dimensional techniques of [15] cannot apply.

- Extending such a 2D Kirchhoff-based Riemann solver, within a Godunov scheme, toward a nonlinear "Pressure-Gradient model" like in [36, Chap. 9] as well.
- At last, establishing a general relation between our WB techniques for hyperbolic equations with so-called "non-standard finite-differences", studied in e.g. [27], may open new possibilities for extensions toward other classes of PDE's.

## References

1. P. Bagnerini, R.M. Colombo, A. Corli, On the role of source terms in continuum traffic flow models. Math. Comput. Model. **44**, 917–930 (2006)
2. B. Bidégaray, J.M. Ghidaglia, Multidimensional corrections to cell-centered finite volume methods for Maxwell equations. Appl. Numer. Math. **44**(3), 281–298 (2003)
3. F. Bouchut, *Nonlinear Stability of Finite Volume Methods for Hyperbolic Conservation Laws, and Well-Balanced Schemes for Sources*, Frontiers in Mathematics series (Birkhäuser, Basel, 2004). ISBN 3-7643-6665-6
4. T. Boukadida, A.-Y. LeRoux, A two-dimensional version of the Lax-Friedrichs scheme. Math. Comput. **63**, 541–553 (1994)
5. M. Brio, A.R. Zakharian, G.M. Webb, Two-dimensional riemann solver for euler equations of gas dynamics. J. Comput. Phys. **167**, 177–195 (2001)
6. B. Despres, Ch. Buet, The structure of well-balanced schemes for linear Friedrichs systems. Appl. Math. Comput. doi:10.1016/j.amc.2015.04.085
7. D. Drikakis, P.K. Smolarkiewiczy, On spurious vortical structures. J. Comput. Phys. **172**, 309–325 (2001)
8. V. Elling, The carbuncle phenomenon is incurable. Acta Math. Sci. **29B**(6), 1647–1656 (2009)
9. V. Elling, Relative entropy and compressible potential flow. Acta Math. Sci. Ser. B Engl. Ed. **35**(4), 763–776 (2015)
10. H. Gilquin, J. Laurens, C. Rosier, Multi-dimensional Riemann problems for linear hyperbolic systems: Part I, eds. A. Donato et al. Nonlinear Hyperbolic Problems: Theoretical, Applied, and Computational Aspects, Friedr. Vieweg & Sohn Verlagsgesellschaft mbH, Braunschweig/Wiesbaden (1993), pp 284–290
11. H. Gilquin, J. Laurens, C. Rosier, Multi-dimensional Riemann problems for linear hyperbolic systems: Part II, eds. A. Donato et al. Nonlinear Hyperbolic Problems: Theoretical, Applied, and Computational Aspects, Friedr. Vieweg & Sohn Verlagsgesellschaft mbH, Braunschweig/Wiesbaden (1993) pp. 284–290
12. H. Gilquin, J. Laurens, C. Rosier, Multi-dimensional Riemann problems for linear hyperbolic systems. RAIRO—Model. Math. Anal. Num. **30**, 527–548 (1996)
13. S.K. Godunov, Reminiscences about difference schemes. J. Comput. Phys. **153**, 6–25 (1999)
14. L. Gosse, *Computing Qualitatively Correct Approximations of Balance Laws* (Springer, New York, 2013). ISBN 978-88-470-2891-3
15. L. Gosse, *A two-dimensional version of the Godunov scheme for scalar balance laws*, vol. 52. (SIMAI Springer Series, Springer, 2014), pp.626–652
16. L. Gosse, A well-balanced and asymptotic-preserving scheme for the one-dimensional linear Dirac equation, BIT Numer. Math. **55**, 433–458 (2015)
17. L. Gosse, Locally inertial approximations of balance laws arising in (1+1)-dimensional general relativity, SIAM J. Applied Math. **75**(3), 1301–1328
18. S. Jin, Z. Xin, The relaxation schemes for systems of conservation laws in arbitrary space dimension. Comm. Pure Appl. Math. **48**, 235–276 (1995)
19. A. Jüngel, *Transport Equations for Semiconductors*, Lecture Notes in Physics (Springer, Berlin, 2009)

20. L. Hsiao, K. Zhang, The relaxation of the hydrodynamic model for semiconductors to the driftdiffusion equations. J. Differ. Equ. **165**, 315–354 (2000)
21. J. Laurens, Multi-dimensional numerical schemes, eds. A. Donato et al. Nonlinear Hyperbolic Problems: Theoretical, Applied, and Computational Aspects, Friedr. Vieweg & Sohn Verlags-gesellschaft mbH, Braunschweig/Wiesbaden (1993), pp. 393–400
22. D. Li, S. Qian, Solutions for a hydrodynamic model of semiconductors. J. Math. Anal. Appli. **242**, 237–254 (2000)
23. J. Li, M. Lukacova, G. Warnecke, Evolution Galerkin schemes applied to two-dimensional Riemann problems for the wave equation system. DCDS-B **9**, 559–576 (2003)
24. T. Li, W. Sheng, The general Riemann problem for the linearized system of two-dimensional isentropic flow in gas dynamics. J. Math. Anal. Appl. **276**, 598–610 (2002)
25. J. Li, T. Zhang, S. Yang. The Two-Dimensional Riemann Problem in Gas Dynamics, Pitman Monographs and Surveys in Pure and Applied Mathematics (1998). ISBN 0582244080
26. P. Marcati, R. Natalini, Weak solutions to a hydrodynamic model for semiconductors and relaxation to the drift-diffusion equation. Archive Rati. Mech. Anal. **129**, 129–145 (1995)
27. R.E. Mickens, *Applications of Nonstandard Finite Difference Schemes* (World Scientific, Singapore, 2000)
28. S. Mishra, E. Tadmor, Constraint preserving schemes using potential-based fluxes. I. Multidimensional transport equations, Comm. Computational. Physics **9**(3), 688–710 (2010)
29. M.H. Pham, M. Rudgyard, E. S'uli, *Bicharacteristic methods for multi-dimensional hyperbolic systems*, Godunov methods: theory and applications, ed. by E.F. Toro (Kluwer Academic/Plenum, New York) 2001
30. F. Poupaud, M. Rascle, J.P. Vila, Global solutions to the isothermal Euler-Poisson system with arbitrarily large data. J. Differ. Equ. **123**, 93–121 (1995)
31. P.L. Roe, Discontinuous solutions to hyperbolic systems under operator splitting. Numer. Methods Partial Differ. Equ. **7**, 277–297 (1991)
32. P.L. Roe, Vorticity Capturing, AIAA paper 01-2523 (2001)
33. B. Wendroff, A two-dimensional hlle riemann solver and associated Godunov-type difference scheme for gas dynamics. Comput. Math. Appl. **38**, 175–185 (1999)
34. C. Shuxing, Multidimensional Riemann problem for semilinear wave equations. Comm. Partial Differ. Equ. **17**(5–6), 715–736 (1992)
35. Y. Zhang, Global solution to a cubic nonlinear Dirac equation in 1 + 1 dimensions, arXiv:1304.1989vl [math.AP] 7 Apr 2013
36. Y. Zheng, Systems of Conservation Laws: Two-Dimensional Riemann Problems, Progress in Nonlinear Differential Equations and Their Applications. **38** (Birkhauser Verlag)

# Index




D. Amadori and L. Gosse, *Error Estimates for Well-Balanced Schemes on Simple Balance Laws*, SpringerBriefs in Mathematics,
DOI 10.1007/978-3-319-24785-4

Printed in the United States
By Bookmasters